Ergebnisse der Physiologie · Reviews of Physiology

Ergebnisse der Physiologie
Biologischen Chemie und experimentellen Pharmakologie

Reviews of Physiology
Biochemistry and Experimental Pharmacology

62

Herausgeber/Editors

L. Brown, Oxford · H. Holzer, Freiburg · R. Jung, Freiburg
K. Kramer, München · O. Krayer, Boston · F. Lynen, München
P. A. Miescher, Genf · W. D. M. Paton, Oxford
H. Rasmussen, Philadelphia · A. E. Renold, Genf
U. Trendelenburg, Würzburg · H. H. Weber, Heidelberg

Mit 14 Abbildungen

Springer-Verlag Berlin Heidelberg GmbH 1970

ISBN 978-3-662-31102-8   ISBN 978-3-540-36192-3 (eBook)
DOI 10.1007/978-3-540-36192-3

Library of Congress Catalog Card Number 68-37142.

Titel-Nr. 4782

Inhalt

Mitarbeiter

Bergel, Franz, Prof. Dr., Magnolia Cottage, Bel Royal, Jersey, C. I./England

Marshall, Jean M., Dr., Division of Biological and Medical Sciences, Brown University, Providence, Rhode Island 02912/USA

Thomitzek, W.-D., Dozent Dr., Chemisches Laboratorium des Stadtkrankenhauses, DDR-7033 Leipzig, Friesenstr. 8

Vogt, Marthe, Dr., Acricultural Research Council, Institute of Animal Physiology, Babraham, Cambridge/England

J H Gaddum

John Henry Gaddum, 1900-1965

MARTHE VOGT

It is now four years ago since physiology and pharmacology lost one of their most eminent figures in Sir JOHN GADDUM. His death was caused by malignant disease, knowingly and uncomplainingly borne, and fought with exemplary energy and courage. Reluctantly he resigned from the Directorship of the Agricultural Research Council's Institute of Animal Physiology (Babraham, Cambridge) a few months before his death, but even after his resignation he finished a monograph he was writing on the neurological basis of learning [1].

GADDUM was born in Cheshire, the eldest of six children; his father, HENRY EDWIN GADDUM, stirred in him an interest in natural history; this interest never left him and embraced all aspects of nature, animate and inanimate. His father's family had some German ancestors and a few relations still live in the Ruhr.

In 1929, GADDUM married IRIS MARY HARMER, a fellow medical graduate who had been a brilliant medical student and who became (and still is) a consultant dermatologist. They shared many interests during a long life of deep affection and mutual support. Their three daughters are married. Although GADDUM had been bed-ridden for weeks, he had the last great joy of being able to get up and attend the wedding of his youngest daughter, Phyllis, to the Cambridge dermatologist ROBERT H. CHAMPION.

GADDUM was educated at Rugby and Trinity College, Cambridge, where he read first mathematics and later physiology. He completed his medical studies at University College Hospital, London, from where he qualified as a doctor in 1925. It was almost accidental that his first post was with J. W. TREVAN at the Wellcome Physiological Research Laboratories in Beckenham. Yet TREVAN's interest in obtaining quantitative data in pharmacology could not have been more akin to GADDUM's interests and gifts. GADDUM's paper from Beckenham [2] on "The action of adrenaline and ergotamine on the uterus of the rabbit" (1926) laid the foundation of much future work by him [3, 4] and others on drug antagonism. In 1927, GADDUM moved to the National Institute for Medical Research (Hampstead, London) where H. H. (later Sir HENRY) DALE was looking for a co-worker. During the next seven years GADDUM's two main lines of interest became apparent: the application of mathematics to make pharmacology an exact and quantitative discipline [5—8], and the search for tissue constituents with dynamic rather than static properties, i.e. endogenous substances which acted on smooth muscle, secretion and last, but not least, nervous activity and thus were in control of most physiological events. Such substances, of which histamine [9—12], acetylcholine, substance P [13—16], catecholamines [17—19], kinins [20] and 5-hydroxytryptamine are but examples, can initially only be detected by their biological actions. This explains the time and energy Gaddum devoted throughout the years to the perfection of bioassays, to improvement of their accuracy, their specificity and their sensitivity [21—25]. Such work started in Hampstead, but was still successfully pursued during Gaddum's last years in Babraham; there he invented (in 1960) the push-pull cannula [26] to obtain and assay, without delay, labile active materials released by brain tissue; there (1926) he found a sensitive test for uridine diphosphate in the goldfish intestine [27] suspended in a micro-bath; in fact, Gaddum's last experimental paper (1964) [28] is the description of "an improved microbath". In addition to the work on bioassays, both experimental and mathematical, two discoveries of great importance for the future of physiology were made during the time Gaddum spent in Hampstead: with Schild [29] he devised the first estimation of adrenaline by the fluorescence developed when catecholamines are treated with alkali in the presence of oxygen; with Feldberg [30] he showed that acetylcholine was the transmitter at ganglionic synapses.

The years 1934 to 1958 saw Gaddum, who considered himself (and was) a physiologist, as Professor of Pharmacology in Cairo, at University College, London, at the College of the Pharmaceutical Society, London and, from 1942 onwards, at Edinburgh. He was interested in teaching, and his textbook on Pharmacology, first published in 1940, saw five editions and was translated into many languages. Its characteristic, concise style, the stress on essentials, the clear illustrations and humorous asides, have won it many admirers. Its briefness makes it essential to pay attention to every word, a fact that some students fail to realize.

Edinburgh was the only university in which he stayed long enough to make his laboratory into the pharmacological Mecca it became. He had many British and foreign students and co-workers on whose work he made an indelible mark. His kindness endeared him to his pupils and staff; his critical mind was the more admirable because he lacked prejudice completely and was willing to listen to any idea or suggestion, however unorthodox. His integrity and sense of duty were beyond praise.

Work that came out during the Edinburgh period concerned the identification of noradrenaline as the transmitter at adrenergic synapses, the functions of histamine, of substance P, and of bradykinin. Special mention should be made of the work on 5-hydroxytryptamine; with his colleagues he mapped out its distribution in the brain [31]; he discovered that its most potent antagonist on the rat uterus was lysergic acid diethylamide [32]; this immediately suggested to GADDUM that the hallucinatory effect of the amide might be due to an action antagonizing that of cerebral 5-hydroxytryptamine. He later revised this simple hypothesis in the light of further evidence [33], but the idea of the importance of 5-hydroxytryptamine in brain function is daily gaining ground in work that had its first impetus from GADDUM's findings.

GADDUM's last seven years were spent as director of the Agricultural Research Council's Institute of Animal Physiology in Babraham, Cambridge. Always ready to accept a new intellectual challenge, the importance of this administrative post and the necessity to familiarize himself with problems on agriculture attracted him; so did Cambridge, where he had lived through happy student days and where his wife had family ties. In preparation for his new task he visited agricultural centres and laboratories of animal physiology all over the world in order to learn what was done elsewhere. During his Directorship, size and scientific output of the Institute grew from year to year, a witness to his successful leadership. His own interests were more and more centred on the nervous system and its transmitter substances [34]; as far as a busy life permitted, he took part in experiments in which substances released from brain tissue were being collected and identified, and in attempts at separating into its components the polypeptide substance P [35—37], discovered by him and VON EULER in 1931 [13], and one fraction of which has interesting central actions.

GADDUM was a great traveller, lectured in many lands, visited laboratories in many continents and gave advice to many scientists. Of the large number of honours conferred on him three gave him particular pleasure — the Fellowship of the Royal Society in 1945, the Knighthood conferred on him in 1964, and the honorary LL.D. given him in 1965 by Edinburgh University.

References

For further biographical details see W. FELDBERG: John Henry Gaddum. Biograph. Mem. Fellows roy. Soc. 13, 57—77 (1967).

1. 1965. The neurological basis of learning. In: Perspect. Biol. Med. 8, 436—474.
2. 1926. The action of adrenalin and ergotamine on the uterus of the rabbit. J. Physiol. (Lond.) 61, 141—150.
3. 1943. Introductory address. Part I. Biological aspects: the antagonism of drugs. Trans. Faraday Soc. 39, part 12, 323—332.
4. 1957. Theories of drug antagonism. Pharmacol. Rev. 9, 211—268.
5. 1933. Reports on biological standards. III. Methods of biological assay depending on a quantal response. Med. Res. Council Spec. Rep. Ser., No 183.
6. 1953. Bioassays and mathematics. Pharmacol. Rev. 5, 87—134.

7. 1953. Review article. Simplified mathematics for bioassays. J. Pharm. Pharmacol. **6**, 345—358.

8. 1945. Lognormal distributions. Nature (Lond.) **156**, 463.

9. 1930. (With R. B. Bourdillon and R. G. C. Jenkins.) The production of histamine from histidine by ultra-violet light and the absorption spectra of these substances. Proc. roy. Soc. B **106**, 388—398.

10. 1935. (With G. S. Barsoum.) The pharmacological estimation of adenosine and histamine in blood. J. Physiol. (Lond.) **85**, 1—14.

11. 1935. (With G. S. Barsoum.) The liberation of histamine during reactive hyperaemia. J. Physiol. (Lond.) **85**, 13 P.

12. 1936. (With G. S. Barsoum.) The effect of cutaneous burns on the blood-histamine. Clin. Sci. **2**, 357—362.

13. 1931. (With U. S. v. Euler.) An unidentified depressor substance in certain tissue extracts. J. Physiol. (Lond.) **72**, 74—87.

14. 1961. The estimation of substance P in tissue extracts. From Symposium on Substance P, p. 7—14. Sarajevo (ed. P. Stern). Scientific Society of Bosnia and Herzegovina, Yugoslavia.

15. 1961. (With J. Cleugh, P. Holton and E. Leach.) The assay of substance P on the fowl rectal caecum. Brit. J. Pharmacol. **17**, 144—158.

16. 1961. (With J. C. Szerb.) Assay of substance P on goldfish intestine in a microbath. Brit. J. Pharmacol. **17**, 451—463.

17. 1939. (With C. S. Jang and H. Kwiatkowski.) The effect on the intestine of the substance liberated by adrenergic nerves in a rabbit's ear. J. Physiol. (Lond.) **96**, 104—108.

18. 1939. (With H. Kwiatkowski.) Properties of the substance liberated by adrenergic nerves in the rabbit's ear. J. Physiol. (Lond.) **96**, 385—391.

19. 1947. (With L. G. Goodwin.) Experiments on liver sympathin. J. Physiol. (Lond.) **105**, 357—369.

20. 1958. (With E. W. Horton.) The extraction of human urinary kinin (substance Z) and its relation to the plasma kinins. Br. J. Pharmac. **14**, 117—124.

21. 1928. (With J. W. Trevan, Ellen Boock and J. H. Burn.) The pharmacological assay of digitalis by different methods. Quart. J. Pharm. **1**, 10—26.

22. 1928. (With U. G. Bijlsma and J. H. Burn.) A comparison of the oxytocic, pressor and anti-diuretic activities of commercial samples of pituitary extract. Quart. J. Pharm. **1**, 493—508.

23. 1949. (With W. S. Peart and M. Vogt.) The estimation of adrenaline and allied substances in blood. J. Physiol. (Lond.) **108**, 467—481.

24. 1953. The technique of superfusion. Brit. J. Pharmacol. **8**, 321—326.

25. 1933. (With H. C. Chang.) Choline esters in tissue extracts. J. Physiol. (Lond.) **79**, 255—285.

26. 1961. Push-pull cannulae. J. Physiol. (Lond.) **155**, 1—2P.

27. 1963. (With M. W. Smith.) A pharmacologically active substance in mammalian tissue extracts. Proc. roy. Soc. B **157**, 492—506.

28. 1964. An improved microbath. Brit. J. Pharmacol. **23**, 613—619.

29. 1933. (With H. Schild.) A sensitive physical test for adrenaline. J. Physiol. (Lond.) **80**, 9P.

30. 1934. (With W. Feldberg.) The chemical transmitter at synapses in a sympathetic ganglion. J. Physiol. (Lond.) **81**, 305—319.

31. 1954. (With A. H. Amin and T. B. B. Crawford.) The distribution of substance P and 5-hydroxytryptamine in the central nervous system of the dog. J. Physiol. (Lond.) **126**, 596—618.

32. 1953. Antagonism between lysergic acid diethylamide and 5-hydroxytryptamine. J. Physiol. (Lond.) **121**, 15 P.

33. 1956. (With M. Vogt.) Some central actions of 5-hydroxytryptamine and various antagonists. Brit. J. Pharmacol. **11**, 175—179.

34. 1962. Chemical transmission in the central nervous system. In: Frontiers in brain research, p. 165—190. New York: Columbia University Press.

35. 1963. (With Joan Cleugh.) The stability of purified preparations of substances P. Experientia (Basel) **19**, 72.

36. 1964. (With Mirjana Randić and M. W. Smith.) An antistrychnine extract from horse intestine. J. Physiol. (Lond.) **172**, 207—215.

37. 1964. (With Joan Cleugh, A. A. Mitchell, M. W. Smith and V. P. Whittaker.) Substance P in brain extracts. J. Physiol. (Lond.) **170**, 69—85.

Adrenergic Innervation of the Female Reproductive Tract: Anatomy, Physiology and Pharmacology

Jean M. Marshall*

With 2 Figures

Table of Contents

* Division of Biological and Medical Sciences, Brown University, Providence, R. I. 02912.

I. Introduction

The last general review of the autonomic innervation of the reproductive tract was made by GRUBER in 1933. This was a comprehensive work which encompassed the sympathetic and parasympathetic innervation of the urogenital systems of both the male and female, and its text and references give an excellent survey of the early literature. In the 36 years since GRUBER's review many major advances have taken place in anatomy, physiology, biochemistry and pharmacology as newer, more refined and more sophisticated techniques became available. These advances provide information which is pertinent to our knowledge and understanding of the autonomic innervation of the reproductive organs. It therefore seemed timely to appraise some of the recent findings in the hope that they will contribute to a better understanding of the role of the autonomic nervous system in reproduction. The present review, unlike that of GRUBER, is more circumscribed and considers only the adrenergic innervation of the female reproductive tract, the ovaries, fallopian tubes, uterus and vagina, with the primary emphasis placed upon peripheral, adrenergic mechanisms. The review is selective not exhaustive, and the topics and references chosen reflect the interests of the author.

It is commonly stated that transection of the spinal cord and/or peripheral nerves going to the internal reproductive organs in the female has apparently a minor effect on copulation, implantation, pregnancy and parturition in many species including rat, rabbit, cat, monkey and human (REYNOLDS, 1965; ANDERSON et al., 1963). This is sometimes taken to mean that the extrinsic nerves are relatively unimportant to the normal functioning of the female reproductive organs. However, the fact that an organ can function in the *apparent* absence of nervous influences does not diminish the importance of such influences on the regulation and modulation of functional activity. Morphological and recent histochemical evidence indicates that many of the adrenergic fibers innervating the internal reproductive organs originate in ganglion cells located near or in these organs. The short, post-ganglionic axons arising from these terminal ganglia would not necessarily degenerate when the spinal cord is sectioned or when the presacral nerves are cut. Thus some degree of adrenergic activity could still be retained at the local peripheral level. Both the catecholamine content and the density of adrenergic nerve fibers in various portions of the female reproductive tract are altered in some species during pregnancy and during estrogen domination (SJÖBERG, 1967) suggesting that the amount and distribution of neurotransmitter liberated during nerve stimulation depends upon the hormonal state of the individual. The well-known change in response of the cat uterus to hypogastric nerve stimulation from excitation in the gravid uterus to inhibition in the non-gravid (originally noted by CUSHNY and by DALE in 1906) is not

due to the presence of specific inhibitory and excitatory fibers in the hypogastric nerve, or to a change in the nature of the neurotransmitter during
pregnancy as was previously suggested (see GRUBER, 1933). Instead, the
"receptor" sites on the effector (myometrial) cells themselves are altered in
some manner which is related to their hormonal environment, so that the
same neurotransmitter (norepinephrine) is excitatory during pregnancy and
inhibitory in other circumstances. All of these findings, which will be discussed
in more detail later, point to the possibility that peripheral adrenergic
mechanisms may influence reproductive functions in a more subtle manner
than previously recognized. Furthermore, the effects of these nervous influences
may depend upon the hormonal state of the animal.

This review begins with an account of the anatomy and histology of the
extrinsic and intrinsic adrenergic nerve pathways to the internal reproductive
organs. Included in this section is a description of the regional distribution
of the intrinsic nerves as detected by fluorescence microscopy and of the
nerve-muscle relations as revealed by electron microscopy. The next section
considers the catecholamine content of the reproductive organs emphasizing
the correlations between the biochemical and histochemical techniques and
the neuroendocrine interactions. The third section deals with the effects of
extrinsic nerve stimulation and begins with a summary of some salient points
about the electrophysiology of autonomic fibers in general and the hypogastric nerves and uterine nerves in particular. The review ends with a discussion of the actions of epinephrine and norepinephrine on the peripheral
effector organs at the cellular level.

II. Anatomy

A. Extrinsic innervation

The first critical and one of the most detailed descriptions of the innervation of the female internal genital organs is that of FRANKENHÄUSER (1867).
He correlated the work of previous investigators and, in addition, presented
his own gross and microscopic findings with particular emphasis on the human.
The origin and distribution of the pelvic, autonomic nerves in the rabbit and
cat were described in detail by LANGLEY and ANDERSON in their classical
series of papers in the Journal of Physiology 1894—1896. The results of their
meticulous experiments form the basis of the modern descriptions of these
nerves in domestic and laboratory animals. These and other important contributions are contained in the excellent reviews of DAVIS (1933), KRANTZ
(1959) and REYNOLDS (1965) and can be summarized as follows.

The pelvic adrenergic nerves arise in the lumbar region (1st to 6th segments) of the spinal cord and exit in the white rami communicantes to the
4th to 6th lumbar ganglia of the sympathetic chain. Nerve bundles run from

the lumbar vertebral ganglia to the inferior mesenteric ganglia located at the bifurcation of the aorta near the inferior mesenteric artery. In the human, the abdominal components of the pelvic adrenergic nerves are the superior, the middle and the inferior hypogastric plexuses. The superior hypogastric plexus begins just below the inferior mesenteric artery and is composed of one to three intercommunicating nerve bundles connected with the inferior mesenteric ganglia and with the lumbar sympathetic ganglia. The superior hypogastric plexus merges into the middle hypogastric plexus which then divides at the level of the first sacral vertebra into several branches going to the right and left sides of the pelvis. These form the right and left inferior hypogastric plexus each of which descends the pelvis in a posterior position then curves laterally and finally enters the sacrouterine fold or ligament. There, on either side of the uterus at the base of the broad ligament and close to the uterine cervix, they merge into the pelvic plexus or, as it is sometimes called, Frankenhäuser's plexus. The pelvic plexus contains, in addition to the branches of the inferior hypogastric plexus, many ganglion cells and nerve ramifications including the visceral branches of the 2nd, 3rd, and 4th sacral (cholinergic) nerves, the nervi erigentes. All pelvic viscera are supplied almost exclusively from the pelvic plexus. Whether any specific component of the plexus supplies one area more than another is not known. The greater part of the plexus divides further into large branches that enter the uterus in the region of the internal os, while the smaller part supplies the vagina and bladder. The branches of the plexus that supply the uterus enter the isthmus area, primarily the sacrouterine fold or ligament, and run to the uterus and fallopian tubes via the broad ligament. The uterus also receives a small part of its innervation directly from the inferior hypogastric plexus.

In the human, the hypogastric nerves are contained in a series of plexuses; in the cat, rabbit, dog, rat and guinea pig, on the other hand, two discrete hypogastric nerves arise from the inferior mesenteric ganglia. This fortuitous situation is advantageous to the physiologist or pharmacologist wishing to study, for example, the effects of nerve stimulation on the pelvic viscera or the electrophysiologic properties of an autonomic nerve which is relatively easy to isolate. Each hypogastric nerve divides into a dorsal and a ventral branch. The dorsal branch runs to the pelvic plexus, while the ventral branch joins ganglion formations in the dorso-lateral wall of the vagina.

LANGLEY and ANDERSON (1894) showed that, in the cat and rabbit, the hypogastric nerve contains both non-myelinated and myelinated fibers. They also pointed out (1895 b) that a proportion of the non-myelinated fibers are preganglionic and are interrupted in ganglia situated in or near the organ innervated, such ganglia being more frequent in the rabbit than in the cat. They claimed that the ratio of pre- to postganglionic fibers in the hypogastric nerve varied from animal to animal. Recently VANOV and VOGT (1963) have shown that the hypogastric nerve in the cat contains nerve cells along its entire length. It is now generally acknowledged that the preganglionic fibers of the pelvic adrenergic nerves may synapse with their postganglionic neurons in a variety of locations including the lumbar vertebral ganglia, the inferior mesenteric ganglia, ganglia in the pelvic plexus, ganglia in or near the pelvic viscera.

The innervation of the ovary and oviduct have received much less attention than that of the uterus. According to MITCHELL (1938), the ovarian nerves originate from three sources. The superior ovarian nerves come from the intermesenteric nerves and from the renal plexus and descend along the outer side of the ovarian blood vessels to the ovary. The middle ovarian nerves are usually paired and also supply the fallopian tubes and the utero-tubal junction. The inferior ovarian nerves may be three or four in number and come from the inferior hypogastric plexus or the lower end of the hypogastric nerve. They also supply the oviduct. In fact, it has been suggested that, in cats and rabbits, the entire adrenergic nerve supply to the oviduct arises from the inferior ovarian nerves which in these species branches from the lower end of the hypogastric nerve (LANGLEY and ANDERSON, 1895a; BRUNDIN, 1965). About half of the ovarian nerves are post-ganglionic with their cell bodies in the inferior mesenteric or spinal ganglia. The remainder are pre-ganglionic and synapse in ganglia located in or near the ovaries and oviduct (LANGLEY and ANDERSON, 1895a and b; BRUNDIN, 1965). Hence, as with the uterus, at least part of the adrenergic innervation of the ovary and oviduct arises from ganglia situated *in* or *near* the effector organs.

B. Intrinsic innervation

1. Light microscopy

Classical histological techniques have contributed considerably to our knowledge of the general morphology and topography of the intrinsic nerve supply to the internal reproductive organs. However, these techniques do not distinguish between adrenergic and other types of nerve fibers in the tissue, nor do they provide detailed information about the relations between nerve terminals and the effector cells. The two modern methods that have been profitably employed to study these two problems are fluorescence and electron microscopy. Before proceeding to a consideration of the recent studies utilizing these methods, some of the earlier investigations with light microscopy will be mentioned since they provide the background for the more recent work.

Much of this early work was concerned primarily with the intrinsic nerves in the uterus. Both myelinated and unmyelinated nerves enter the uterus along the blood vessels. Within the body of the uterus, large bundles of myelinated and unmyelinated nerve fibers penetrate the muscular layers then branch repeatedly and run parallel to the muscle bundles. The myelinated fibers lose their myelin sheaths and enter a terminal reticulum which forms the ultimate link between nerve fibers and uterine muscle cells (KRANTZ, 1959; JACOBSON and NIEVES, 1961). The smooth muscle in the walls of the blood vessels within the uterus also receive a rich supply of predominantly unmyelinated fibers (KRANTZ, 1959). The precise nature of the relationship between the terminal nerve network and the myometrial cells could never have been discerned with the light microscope. One of the reasons for this is the peculiar resistance of the intercellular space in both the myometrium and endometrium to the diffusion of staining fluids. The resistance is due to the presence of viscous elements such as hyaluronic acid in the connective tissue. This technical difficulty was at least partially overcome by PALLIE et al. (1954) who injected hyaluronidase locally into the uterus of the rabbit. After treatment with hyaluronidase both the methylene blue and Golgi stains penetrated the tissues much more readily than in untreated animals. The results of this careful study are among the most enlightening of all investigations on intrinsic uterine nerves visualized by light microscopy. PALLIE et al. (1954) noted that the larger intrinsic nerve fibers were composed of plexiform strands, each strand containing from 12 down to 2 or 3 individual fibers, myelinated and unmyelinated and invested by Schwann cells. At intervals the strands divided and ultimately gave off single fibers which ran parallel to the muscle bundles. Although the exact relation between the individual nerve terminations and the myometrial cells was not obvious, fine nerve branches were observed to lie in the interstices between the muscle elements. These fine nerve fibers were sometimes pointed and sometimes bulbar, and although there was a close spatial approximation of nerve to muscle fibers there was no suggestion of a specific nerve-muscle connection as at the motor end plate in skeletal muscle. In the rabbit uterus the nerve "terminations" were too few to provide each individual muscle cell with a nerve ending, hence PALLIE et al. suggested that stimulation of the motor nerves to the rabbit uterus would produce diffuse rather than precisely localized or controlled activity as is the case with skeletal muscle. These findings have been substantiated and extended to include the guinea-pig uterus by CLEGG (1962).

Nerves have been demonstrated to pass through the myometrium and enter the endometrium penetrating at least as far as the superficial glandular portion in the human, the rhesus monkey and the guinea pig (KRANTZ, 1959; JACOBSON and NIEVES, 1961). KRANTZ (1959) found an extrinsic plexus of myelinated and unmyelinated fibers following the blood vessels into the endometrium and apparently terminating in the stroma and adventitia at the origins of the basilar and spinal arteries. Although he claimed that the outer two-thirds of the human myometrium was devoid of nerves, a more recent study in the rhesus monkey by JACOBSON and NIEVES (1961) using the staining procedures of Pallie et al. (1954) demonstrated the presence of nerve fibers in the superficial glandular portion during the intermenstrual stage. This finding raises the interesting question of the fate of these nerve fibers when the superficial part of the endometrium is shed during menstruation. The fibers must be damaged in the process of sloughing of the endometrial tissue and must undergo subsequent regeneration during the succeeding phase of the cycle. Here, then, might be an example of cyclical, physiological degeneration and regeneration of peripheral nerve fibers which would lend itself to an experimental analysis at the ultrastructural level. The results of such a study may contribute to an overall understanding of the regeneration of peripheral nerve fibers. Also, the presence of nerves, which Jacobson and Nieves (1961) presumed to be postganglionic autonomics, in the endometrium indicates that the nervous system may exert some direct control over the function of the glandular portion of the primate uterus.

2. Fluorescence microscopy

About ten years ago, Eränko (see his 1967 review) observed that, following exposure of sections of the adrenal medulla to formaldehyde solution, a fluorescence developed in some cells. Chemical analysis showed that these cells contained norepinephrine. Since it was known that norepinephrine when treated with formaldehyde forms a fluorescent compound, treatment with formaldehyde was proposed as a histochemical method for the cellular localization of norepinephrine. However, the method, as originally proposed, was not sufficiently sensitive to localize norepinephrine in small adenergic nerve fibers. Then in 1962, Falck and his associates (Falck et al., 1962; Falck, 1962) noted that a very intense fluorescence appeared when monoamines in freeze-dried tissues were exposed to formaldehyde vapor. A successful modification of Eränko's original technique was thereby developed which was sensitive enough to detect catecholamines in nerve endings and in ganglion cells. Catecholamines condense with formaldehyde to form tetrahydroisoquinolines which give a green emission spectrum at 4,800 Å, while fluorescent compounds obtained from 5-hydroxytryptamine are yellow with an emission maximum at 5,300 Å. These spectral differences make it possible to discriminate between catecholamines and 5-hydroxytryptamine within tissues. In addition, epinephrine being a secondary amine is more slowly condensed with formaldehyde than is dopamine or norepinephrine and, hence, can be distinguished from them (Falck, 1962). Along with these advances in histochemistry came the development of sensitive chemical methods for the analysis of tissue catecholamine content (Shore and Olin, 1958; Bertler et al., 1958). Therefore the richness of adrenergic innervation determined by fluorescence microscopy can be checked by parallel chemical analyses of tissue catecholamine content. Histochemical techniques apparently can detect epinephrine and norepinephrine to within 5 per cent of the total tissue concentration as determined chemically (CARLSSON et al., 1957). These two techniques, histochemical and chemical, triggered an explosion of research which has contributed a great deal to our knowledge of the adrenergic innervation of autonomic effector organs.

Before proceeding to a more detailed account of the pattern and distribution of adrenergic nerves within the female reproductive tract, some of the general characteristics of adrenergic neurons as revealed by fluorescence microscopy will be noted briefly. These characteristics concern primarily the intraneuronal distribution of catecholamines and the morphology of the nerve-muscle relationships at the cellular level (FALCK, 1962; NORBERG and HAMBERGER, 1964; MALMFORS, 1965). One of the most striking features of the terminal portions of the adrenergic nerve bundles running within the tissues are the elongated, bead-like varicosities seen along the entire length of the terminal axons (cf. Fig. 1). The varicosities show an intense green to green-yellow fluorescence, and are particularly numerous in those portions of the axons in close contact with the effector cells. Chemical analyses of the norepinephrine content of various tissues, when related to the size and distribution of the bead-like varicosities in the terminal axons visualized directly by fluorescence microscopy, provide indirect evidence that the neurotransmitter is localized in the varicosities. The more intense the fluorescence, the greater the concentration of the catecholamines. Thus the intensity of the fluorescence is qualitatively related to the concentration of norepinephrine in the nerves. There is also good evidence that the concentration of norepinephrine is many times higher in the terminal axons than in other parts of the axon or cell body (FALCK, 1962). The intimate relations between the effector cells and the fine nerve twigs showing varicosities justify the assumption that these fine nerve ramifications are the regions where the transmitter (norepinephrine) is released.

Furthermore, since these bead-like varicosities are distributed all along the terminal axons, as sort of "synapses *en passage*", one axon may be able to affect many effector cells (see Section III.B.2 for additional discussion of this concept).

Much of our present knowledge about the adrenergic innervation of the female internal genital organs (ovary, oviduct, uterus and vagina) utilizing the combined fluorescence microscopy and chemical analysis of tissue catecholamines comes from the work of a group of Swedish investigators (see SJÖBERG, 1967). They have examined a variety of species including rat, rabbit, cat, guinea pig and human. From the results of these studies at least three generalizations can be made: first, the density of adrenergic innervation may vary from organ to organ; second, the delayed, green fluorescence, characteristic of epinephrine is never prominent in any of the tissues examined, suggesting that relatively little epinephrine is present; third, the pattern of distribution of fluorescent nerve fibers within any one organ is principally the same in all species. A more detailed description of the innervation of the individual organs follows.

a) Ovary

Thick bundles of preterminal nerve fibers containing no varicosities and very little fluorescence enter the ovary along with the blood vessels in the hilar region. In the ovary itself, catecholamine-containing nerves are seen in close proximity to blood vessels and also within the stromal fibromuscular layer. The density of this adrenergic nerve network is particularly striking in the cat, human and monkey (ROSENGREN and SJÖBERG, 1967; OWMAN et al., 1967; JACOBOWITZ and WALLACH, 1967) and suggests a possible influence of these fibers on ovarian function. The nature of this influence may be a physical effect on ovarian tone to either aid or retard the expulsion of ova. Many nerve fibers are present in the fibromuscular tissue in the thecal region near the ovarian follicles (JACOBOWITZ and WALLACH, 1967). No unusual anatomical relationship of nerve fibers to follicles is observed. The density of innervation agrees well with the amount of norepinephrine present in the ovaries (ROSEN-GREN and SJÖBERG, 1967).

b) Oviduct

As with the ovary, a portion of the nerves to the oviduct follows along the blood vessels while the remainder is distributed to the smooth muscle layers. The innervation density varies along the course of the oviduct with an increase from the ampulla to the isthmus (BRUNDIN, 1965; SJÖBERG, 1967). At the junction of the ampullary and isthmic portions there is an abrupt increase in density which is confined mainly to the circular muscle layer. This increase in density of innervation combined with a notable thickening of the muscular wall at that region prompted BRUNDIN (1965) to suggest that the isthmus serves as a neurally-controlled sphincter, which may be functionally important in the transport of ova. The number of nerves in the circular layer of smooth muscle then decreases somewhat near the tubo-uterine junction (OWMAN and SJÖBERG, 1966; BRUNDIN, 1965; OWMAN et al., 1967). This pattern of innervation is practically identical in all species studied, although it is sometimes also possible to recognize an enhancement of the muscular innervation at the utero-tubal junction in the cat (ROSENGREN and SJÖBERG, 1967) and in the rat (NORBERG and FREDRICSSON, 1966) but not in the rabbit or human. Hence, in the cat and rat, a sphincter mechanism may operate at the utero-tubal junction as well as at the isthmic portion of the oviduct. The lack of enhancement of adrenergic innervation in the utero-tubal region of the rabbit and monkey is of interest in the light of the suggestion of a sphincter-like action in this region of the oviduct in these animals. Recently,

DAVID and CZERNOBILSKY (1968) reported that the utero-tubal junction in the rabbit contains polyp-like mucosal processes that extend into the lumen and are responsible for the blockade of fluid transport from uterus to tube in this animal. Under these circumstances the need for a smooth muscle sphincter would be minimal.

c) Uterus

In animals possessing a bicornate uterus (e.g. rabbit, cat, guinea pig) the muscular coat of the uterine horns consists of an outer longitudinal coat and an inner circular layer separated by a well developed vascular plexus (SOBOTTA, 1891). Many of the vessels within this plexus are enclosed by green-fluorescent, varicose nerves, but, as in the oviduct (SJÖBERG, 1967), adrenergic nerves were also found to run in close proximity to the myometrial fibers both in the inner circular and outer longitudinal muscle coats. The adrenergic innervation of the cat uterus (ROSENGREN and SJÖBERG, 1967) is richer than that of the rabbit (OWMAN and SJÖBERG, 1966) and the guinea pig (SJÖBERG, 1968). These findings correlate well with the four- to five-fold higher norepinephrine content in the cat uterus (ROSENGREN and SJÖBERG, 1967). In contrast to the cat, rabbit and guinea pig, the adrenergic nerve supply to the rat uterus is limited almost exclusively to the vascular system (NORBERG and FREDRICSSON, 1966). This conspicuous species variation must, of course, be considered in any physiological or pharmacological investigation of adrenergic mechanisms in the myometrium.

The only species examined thus far which shows regional variations in the density of uterine adrenergic innervation is the human (OWMAN, ROSENGREN and SJÖBERG, 1967) where the cervix has the highest innervation density as compared with the corpus or fundus (Fig. 1). It is possible that this distribution of innervation allows a more intricate neural control of the cervix which may be of special importance during delivery and immediately post partum.

The interesting findings of JACOBSON and NIEVES (1961) regarding innervation of the glandular epithelum of the primate uterus were not substantiated by fluorescence microscopy, indicating that the nerves visualized with light microscopy in this portion of the uterus are probably not adrenergic.

d) Vagina

The vagina has a rich adrenergic nerve supply to both the vascular and muscular system in the cat, rabbit, guinea pig and human (SJÖBERG, 1967). Sometimes nerves are seen on the mucosal layer, but here they are usually associated with blood vessels; they were never seen to enter the epithelial layer.

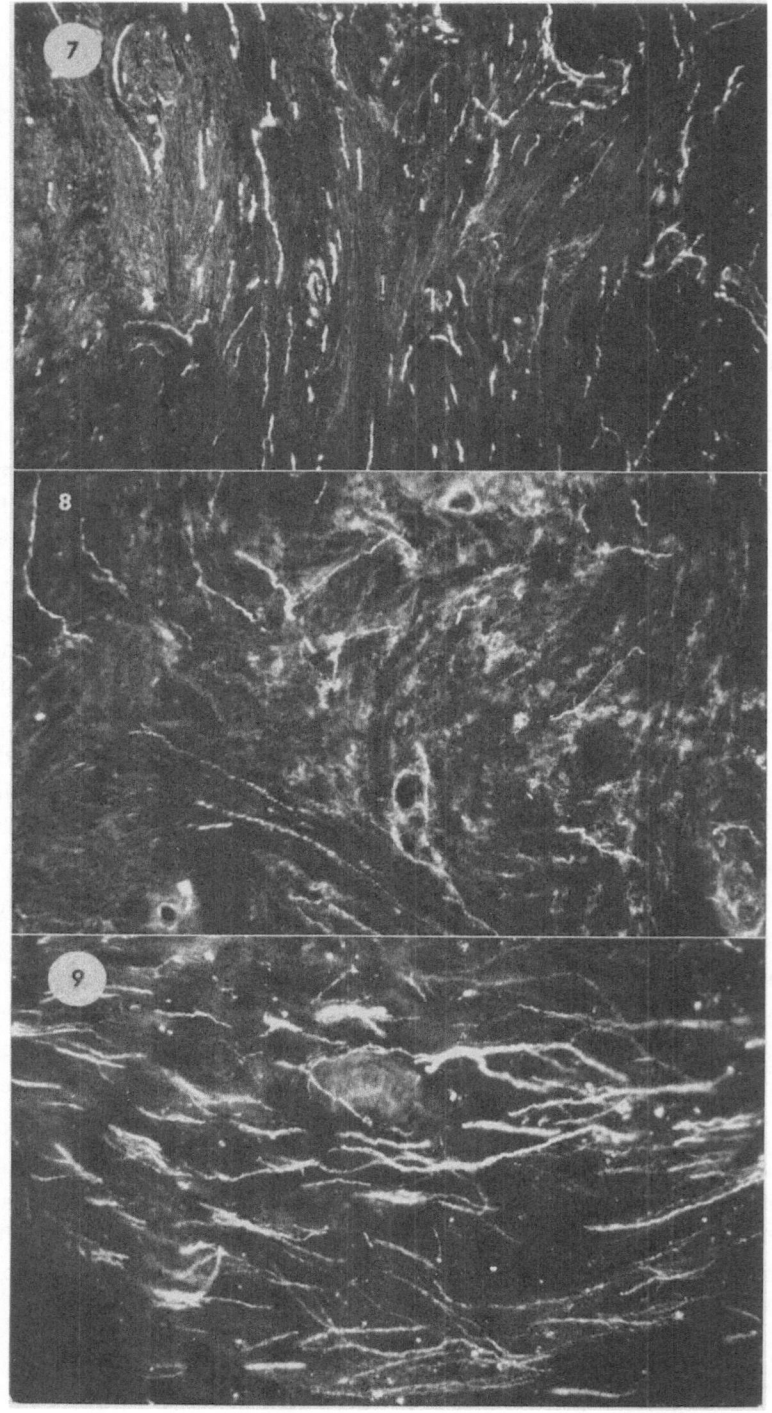

Fig. 1. Adrenergic innervation of human uterus and fallopian tube as visualized by fluorescence histo-chemistry. 7. Section of intramural portion of fallopian tube. Fluorescent nerve terminals are abundant although fewer in number than in the isthmus. Smooth muscles of tube pass over into uterine smooth muscle wall at right half of field. 8. Section of uterine fundus. Only a few fluorescent nerve terminals are present. They are distinguishable from the autofluorescent connective tissue elements by a higher density of fluorescence and varicosed appearance. 9. Section of uterine cervix showing considerable numbers of fluorescent nerves running either singly or in small bundles among the myometrial cells. (All sections × 100.)

(From OWMAN et al., 1967)

e) Peripheral origins of intrinsic innervation

As mentioned earlier, LANGLEY and ANDERSON (1895 b) suggested that at least part of the adrenergic nerves going to the pelvic genitalia were pre-ganglionic and synapsed with ganglion cells located in or near the pelvic viscera. Their evidence for this statement came from studies with nicotine which was found to prevent the contraction of the uterus in response to stimulation of the peripheral end of the severed hypogastric nerve in the rabbit and cat. However, blanching of the uterus was still observed in the presence of nicotine, indicating that the vasomotor component of hypo-gastric stimulation remained intact. Although earlier workers had noted ganglion cells in the pelvic plexus and in the hypogastric nerve, they could not prove that the ganglia were in fact adrenergic relay stations. Recent morphological evidence for the presence of adrenergic ganglia peripheral to the inferior mesenteric ganglia comes from studies utilizing the histochemical and fluorescence techniques (OWMAN and SJÖBERG, 1966; ROSENGREN and SJÖBERG, 1967; OWMAN et al., 1967; SJÖBERG, 1968), and shows that adrenergic ganglion cells are found in the utero-vaginal junction, in the wall of the vagina, and in the connective tissue immediately adjacent to the vagina in rabbit, cat, guinea pig and human. The ganglion formations in these regions contain both fluorescent and nonfluorescent cells, the latter probably being cholinergic. Both types of cells are often enclosed by a network of intensely fluorescent adrenergic nerve terminals containing many varicosities, suggesting a synaptic arrangement involving preganglionic adrenergic nerve terminals. No intrinsic ganglia lying between the layers of smooth muscle cells, similar to those of Auerbach and Meissner in the intestine, have been found in the female genitalia.

The results of denervation experiments support the idea that the utero-vaginal ganglion cells do indeed represent the peripheral adrenergic relays postulated many years ago by LANGLEY and ANDERSON (1895 b). It is well-known that the morphology and transmitter content (and therefore fluores-cence) of postganglionic axons are not usually affected when the preganglionic fibers are cut. But destruction of the postganglionic neurons, or section of the postganglionic axons, results in a rapid loss of transmitter in the post-ganglionic nerve terminals within about 15 hours, and all fluorescence dis-appears within 48 hours (FALCK, 1962; SJÖBERG, 1967). With this information in mind, the Swedish investigators performed various denervation experiments and then, at least one week later, analyzed the changes in fluorescence intensity and in catecholamine content of the internal reproductive organs of rabbits, cats and guinea pigs. Removal of the vertebral ganglia from L_3 to S_2 abolished all fluorescence in the ovarian nerves and in the vascular nerves of the oviduct, uterus and vagina but did not affect the fluorescence in the nerves running within the smooth muscle layers of the latter organs (OWMAN et al., 1966; SJÖBERG, 1967). When the hypogastric nerves were severed, there was only a slight change in the catecholamine content (determined chemically) in the cat oviduct or in the uteri from rabbits and guinea pigs (ROSENGREN and SJÖBERG, 1967; OWMAN et al., 1966; SJÖBERG, 1968), while the catecholamine content was reduced only by about 50 % in the rabbit oviduct and in the cat uterus and vagina. BRUNDIN (1965) noted that division of the hypogastric nerves just below the inferior mesenteric ganglion causes only a partial disappearance of fluorescent nerves and of norepinephrine contents in the rabbit oviduct. These results also support the original suggestion of LANGLEY and ANDERSON (1895 b) that some of the adrenergic nerves running to the pelvic viscera synapse in ganglion cells located in close proximity to the target organs and are therefore not affected by removal of the vertebral ganglia or cutting of the hypogastric nerves. The adrenergic nerves to the blood vessels and to the ovary, however, appear to be mainly postganglionic fibers whose cell bodies are located in the spinal ganglia.

The peripheral adrenergic neurons lying in or near the pelvic viscera have been termed "short adrenergic neurons" by OWMAN and SJÖSTRAND (1965) to distinguish them from the "long adrenergic neurons" arising in the vertebral and inferior mesenteric ganglia. The "short" neurons are believed to represent a special type of ganglion cells, resembling in their location the cholinergic (parasympathetic) ganglia. Thus far they have been found in the urogenital tract of the male as well as of the female (FALCK, 1962; OWMAN and SJÖSTRAND, 1965; SJÖSTRAND, 1965; EL-BADAWI and SCHENK, 1968).

The "short" neurons are, functionally as well as anatomically, different from "long" adrenergic neurons. Norepinephrine containing granules isolated from the vas deferens and vesicular glands of the bull, both structures innervated by "short" adrenergic neurons, have a much slower rate of spontaneous release of norepinephrine than granules isolated from the splenic nerves in that animal (EULER and LISHAJKO, 1966; STJÄRNE and LISHAJKO, 1966). After a single injection of reserpine, the norepinephrine concentration in the uterine adrenergic nerves of the rabbit declines at a much slower rate than that in adrenergic nerves in the heart and spleen (NILSSON, 1964). OWMAN et al. (1966) compared the rate of disappearance and reappearance of norepinephrine in organs innervated by "long" (heart and ovary) versus "short" adrenergic neurons (uterus and vagina) in the rabbit by studying the intensity of fluorescence of the neurons in these organs at varying times after administration of reserpine. The results showed that the transmitter disappeared from the "long" neurons in about 4 hours after a single dose of reserpine while in the "short" neurons transmitter could still be detected as long as 30 hours after injection. These studies may indicate that the rate of spontaneous release of norepinephrine *in vivo* is slower in "short" than in "long" neurons. Recovery of norepinephrine after depletion by reserpine, as indicated by an increase in fluorescence intensity of the adrenergic neurons, occurs earlier and more rapidly in the "short" than in the "long" nerves (SJÖBERG, 1967).

The functional and anatomical differences between these two types of neurons must certainly be taken into account when considering the role of the adrenergic nerves in the regulation or modulation of reproductive processes. Among other things, it is obvious that adrenergic denervation of the oviduct, uterus and vagina in most species cannot be accomplished simply by removal of the lumbar and sacral vertebral ganglia, or by sectioning of the hypogastric nerves, or by the administration of reserpine. Actually it is almost impossible to denervate the pelvic viscera surgically. The complex interconnecting networks of nerve cells and axons in the peripheral ganglia may indicate that local, peripheral reflex mechanisms are involved in the regulation of functional activity of the reproductive organs. As we shall see later (p. 32), the electrophysiological characteristics of these peripheral ganglia and nerve fibers are just beginning to be investigated.

f) Functional changes in neuronal transmitter content; pregnancy

It is possible to estimate changes in the effective density of innervation of the various reproductive organs during pregnancy if the innervation pattern during pregnancy is compared with changes in organ size and weight, i.e. the number and arrangement of fluorescent nerves is correlated with the mass of smooth muscle tissue. ROSENGREN and SJÖBERG (1968) have made such a correlative study in the rabbit and found that the density of adrenergic innervation of the ovary was essentially unchanged during pregnancy. In the oviduct, which shows a considerable increase in weight during pregnancy, the number of fluorescent nerve fibers per amount of smooth muscle remained relatively constant, indicating an increase in total number of nerves which could be visualized by the fluorescence technique. These authors feel that the increase in fluorescence results from an increase in content of transmitter within the individual axons, thus allowing more of them to be visualized by the histochemical technique. This assumption was verified chemically by the finding of an increase in norepinephrine content of the oviduct during pregnancy. In the uterus, on the other hand, there was a decrease in density of adrenergic innervation during the first two weeks of pregnancy and this decline became more marked as pregnancy progressed. Near term, with the exception of the cervix, few if any fluorescent fibers were ever seen in the myometrium. Even in the cervix the innervation density was lower than in the non-pregnant rabbit, as was the intensity of fluorescence of the individual nerve fibers, indicating a reduction of neuronal transmitter content. Similar though less conspicuous changes occurred in the vagina. The number of fluorescent nerves was also markedly reduced in the guinea-pig uterus during the latter part of pregnancy as was the norepinephrine content (SJÖBERG, 1968). This decrease was seen in uterine horns both with and without fetuses, indicating that mechanical factors associated with distention of the uterus during pregnancy were not responsible for the decrease in adrenergic transmitter content and suggesting that hormonal factors might be involved.

To explore the latter possibility, a study was made of the effects of estrogen on the catecholamine content and fluorescence of the neurons in the reproductive tract of the female rabbit (SJÖBERG, 1968). After daily injections of 17-beta-estradiol (0.5 µg/kg) for 14 days the norepinephrine content of the uterus and vagina was increased. This effect was less pronounced in the oviduct (see also, BRUNDIN, 1965) and was totally absent in the ovary and the heart. Despite enlargement of the uterus and vagina with estrogen treatment, the density of the fluorescent nerve terminals was even greater than in untreated animals. The fluorescence of the varicosities of individual axons also appeared more intense, giving them a swollen and distended appearance. Thus, it seemed that estrogen had increased the transmitter content of individual axons allowing previously nonfluorescent axons to be visualized microscopically. During estrogen treatment, the effective innervation density kept pace with the increase in muscle size, just the opposite of what was seen in the uterus at the end of pregnancy. Thus, the changes observed during pregnancy probably result from the combined effects of various hormones including both estrogen and progesterone. The specific effects of progesterone have not as yet been studied.

These alterations in neuronal norepinephrine content during pregnancy and during estrogen treatment are typical only of certain organs in the female reproductive tract and not of adrenergically innervated organs in general (SJÖBERG, 1967). It will be recalled that at least part of the adrenergic innervation of the female internal genitalia is by way of the "short" neurons. SJÖBERG (1967) has made the interesting suggestion that the changes found in the portions of the genital tract during pregnancy and after estrogen treatment may represent another functional distinction between the "short" and "long" adrenergic neurons innervating the pelvic organs.

g) Non-neuronal monoamine-containing cells

In addition to the adrenergic ganglion cells and their peripheral neurons, two other types of monoamine-containing cells have been noticed in the female genital tract: chromaffin cells and cells containing 5-hydroxytryptamine. The chromaffin cells, so-called because they usually give the chromaffin reaction due to their amine content (LEMPINEN, 1964), are small, branching cells found near, within or close to the peripheral adrenergic ganglia in the wall of the vagina in the rabbit, cat and guinea pig (OWMAN and SJÖBERG, 1966; ROSENGREN and SJÖBERG, 1967; SJÖBERG, 1968). The functional significance of these cells is not known, but they have also been observed in relation to other adrenergic ganglia, including the superior cervical ganglion of the cat (NORBERG and HAMBERGER, 1964; NORBERG et al., 1966). It has been suggested (NORBERG et al., 1966) that they may be storage sites for monoamines. The second type of cell has been found in the epithelium of the cat and rabbit vagina and contains 5-hydroxytryptamine (NORBERG et al., 1964; SJÖBERG, 1967). As with the chromaffin cells, the functional significance of the cells containing 5-hydroxytryptamine is unknown.

3. Electron microscopy

Although the literature contains numerous reports about the fine structure of the smooth muscle cells in the female reproductive tract, especially the uterus (GANSLER, 1960; HARMAN et al., 1962; JAEGER, 1962; LAGUENS and LAGRUTTA, 1964) relatively few studies with the electron microscope have been concerned with the innervation of the pelvic viscera. In 1957, CAESAR et al. presented an ultrastructural account of the intrinsic nerves in various smooth muscles including the myometrium, but they published electron micrographs of the urinary bladder only. On the other hand, the literature on the fine structure of the innervation of smooth muscle from other visceral organs (intestine, colon, urinary bladder, vas deferens) with emphasis on the nerve-muscle relations is considerable (BURNSTOCK and HOLMAN, 1966a, for review; BURNSTOCK and ROBINSON, 1967). One recent report, however, concerns the ultrastructure of the nerves in the rat uterus with particular reference to the terminal portions of the nerve axons and their relation to the myometrial cells (SILVA, 1966). This author found that the muscle cell—nerve fiber relationship in the rat myometrium is, in general, similar to that previously noted in other smooth muscles, and can be described as follows: bundles of non-myelinated nerves partly or completely supported by a network of Schwann cells ramify within the myometrium and form the so-called autonomic ground plexus. Flattened portions of the nerve bundles some 10 μm wide are often seen lying parallel to the muscle cells, sometimes passing through indentations in the sarcolemma of two opposing cells. Occasionally single axons (2—3 μm in diameter) partially or completely devoid of their Schwann cell sheaths come to lie within 200 to 300 Å (0.02—0.03 μm) of the smooth muscle sarcolemma (Fig. 2). Since serial sections of this material were not made, it is not possible to state with certainty whether these nerve—muscle contacts are actually nerve terminals or whether the axon emerges from the groove in one muscle cell and proceeds to form close contacts with other muscle fibers. It

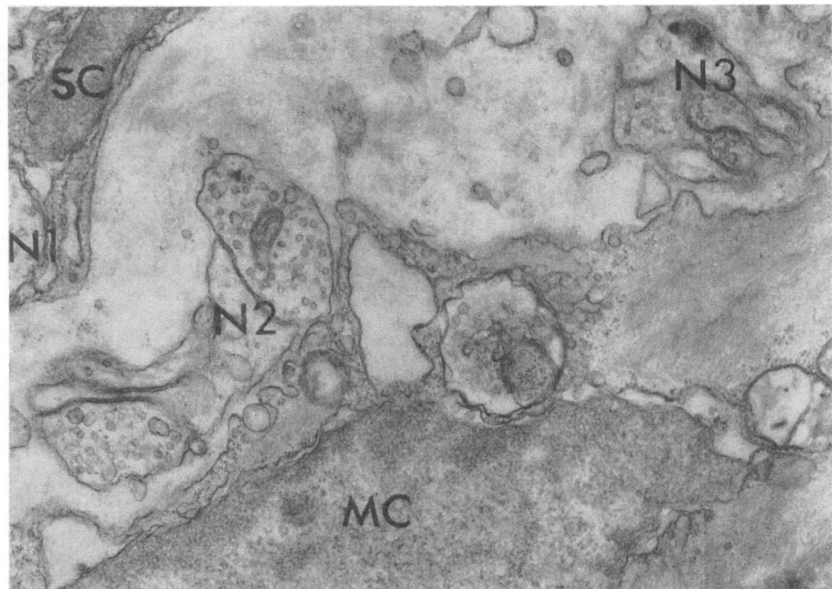

A

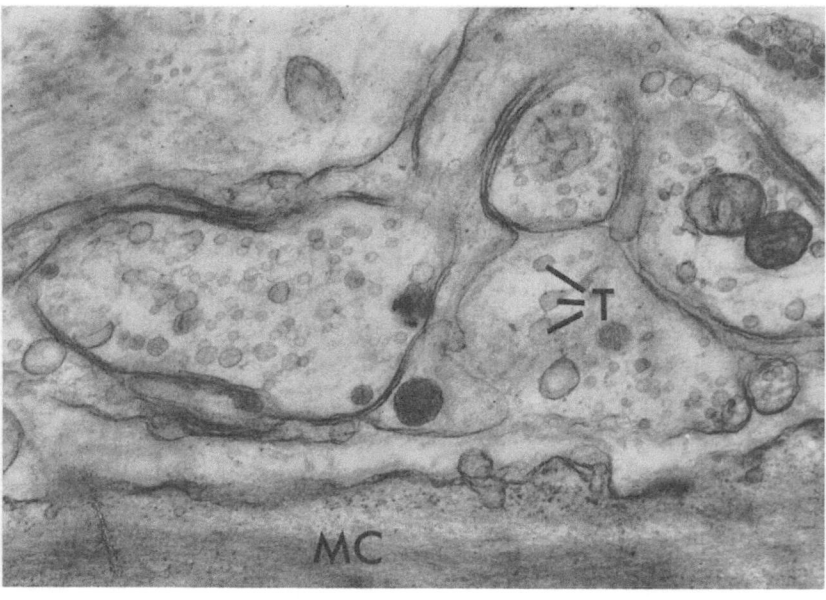

B

Fig. 2A and B. The ultrastructure of the nerve-muscle cell relations in the myometrium of the non-pregnant rat. A Three nerve bundles (N 1, N 2, N 3) can be seen; of these N 2 contains a nerve fiber which is separated from a muscle cell (*MC*) by a gap of about 200 Å (20 nm). The axis cylinders contain many small agranular vesicles and a few small granular vesicles. Schwann cell (*SC*) ×25,000. B Part of a medium sized bundle of nerves lying in close proximity to a muscle cell (*MC*). The agranular vesicles vary in size. Some of the medium granular vesicles contain one or two pale cores. Large vesicle-like structures appear to be associated with a neurotubule (*T*) ×43,000. (From SILVA, 1966) has been suggested that the nerve action potential may release transmitter from many points along the axon as well as from other points in the autonomic ground plexus (BURNSTOCK and HOLMAN, 1966a). In this manner one nerve axon may influence many different smooth muscle cells. Therefore the term "nerve ending" as applied to autonomic neuro-effector junctions including those in the myometrium is best defined in the functional sense to mean that part, or those parts, of the neuron from which the release of transmitter substance occurs during stimulation. Unlike the motor end plate of skeletal muscle, the nerve and smooth muscle cells are not specialized at their points of apposition (BURNSTOCK and HOLMAN, 1966a). However, the axons lying in close juxtaposition to the smooth muscle membrane, as well as those in the autonomic ground plexus some distance away, contain numerous vesicles and mitochondria (BURNSTOCK and HOLMAN, 1966a; SILVA, 1966). Some of the vesicles are granular and others are agranular. The granular ones range in diameter from 300 to 1500 Å (0.03 to 0.15 μm) with the smaller ones containing an electron-dense core. These small granular vescicles may be the storage sites for norepinephrine although this is difficult to verify experimentally (BURNSTOCK and ROBINSON, 1967). The agranular vesicles may contain acetylcholine, or precursors of norepi- and epinephrine, or even some unknown neurotransmitter (BURNSTOCK and HOLMAN, 1966a). Unfortunately, histochemical techniques are not available which enable us to distinguish between adrenergic and cholinergic nerve fibers at the ultrastructural level. In the rat myometrium, at least according to SILVA (1966), the majority of the neuronal vesicles are agranular. It will be recalled that the rat myometrium is notably poor in adrenergic nerve "terminals" as visualized with fluorescence microscopy (NORBERG and FREDRICSSON, 1966).

Electron micrographs indicate that, in the rat, the density of uterine innervation is much less than that of many other smooth muscles. For example, each smooth muscle cell in the vas deferens of the rat and mouse is thought to be closely related to a nerve axon (RICHARDSON, 1962; HOLMAN, 1967). The urinary bladder is likewise richly innervated (CAESAR et al., 1957). The rat myometrium shows a high degree of well-coordinated spontaneous motility of myogenic origin, especially in estrogen-treated animals (BURNSTOCK et al., 1963). This type of activity implies that myogenic cell-to-cell conduction of the action potential is a rapid and efficient means of recruiting muscle fibers into activity. Thus the need for activation of individual muscle cells or bundles via nervous mechanisms is minimal. In line with this suggestion is the report of BERGMAN (1968) that in the estrogen-primed rat uterus there are numerous areas where the sarcolemma of one smooth muscle cell is in close apposition to that of its neighbors. These areas, the so-called "tight junctions", may be the regions of low resistance where the myogenic spread of action current can occur from one cell to another (BURNSTOCK et al., 1963). It has been suggested that the greater the degree of interaction between neighboring smooth muscle cells (i.e. of the myogenic electrical coupling), the lesser the density of innervation (BURNSTOCK and HOLMAN, 1966a). It is unfortunate that the uteri of the rabbit and the cat have not been carefully examined with the electron microscope. In these animals, the adrenergic innervation of the uterus is much more prominent than in the rat (cf. SJÖBERG, 1967), although the uterus is spontaneously active under certain circumstances. A comparative study of the ratio of fine nerve fibers to smooth muscle cells, of the granule content of the nerve axons, of axon-muscle cell relationships and of the relative numbers of "tight junctions" in the uteri of various species might help to clarify the role of the adrenergic nervous mechanisms in uterine motility.

An ultrastructural study of the smooth muscle coat of the rabbit oviduct with special reference to the nerve-muscle relations has recently been reported (KUSHIYA, 1968). In this work, the ampullary wall of the oviduct was used, an area that has been shown by BRUNDIN (1965) and by OWMAN and SJÖBERG (1966) to be poorly innervated by adrenergic nerve fibers. Hence this region of the oviduct is not the best choice of tissue for studies on *adrenergic* innervation. As with the rat uterus, the ampullary region of the rabbit oviduct contains relatively few nerve "endings" (defined as portions of the fine nerve axons containing vesicles). Most of the nerve "endings" are located at some distance (several microns) away from the muscle cells (KUSHIYA, 1968), suggesting that the neurotransmitter would reach the muscle cells by generalized diffusion. The muscle cells are loosely arranged with elements of connective tissue separating individual cells or small bundles of cells. This suggests that myogenic spread of excitation via electrical coupling may be difficult. It is interesting that the spontaneous contractions of the ampulla in the rabbit oviduct are much weaker and less frequent than those of the isthmus (GREENWALD, 1963; BRUNDIN, 1964).

In view of the extremely rich adrenergic innervation of the ovary, especially in the cat and in the human, and its intimate relationship to the developing follicles (JACOBOWITZ and WALLACH, 1967), it would be of interest to investigate the ultrastructure of the fine nerve axons in this organ. Such a study might shed some light on the nature of the nerve-follicular cell interaction.

In sum, electron microscopic techniques, although they do not permit differentiation between adrenergic and cholinergic nerve terminals, nevertheless are very powerful tools for investigating the nerve-effector cell relationships at the ultrastructural level. They have not been fully exploited in studies of the female reproductive organs. Any investigator who is willing to undertake the meticulous, demanding and technically difficult research procedures needed for such studies would be richly rewarded by a harvest of crucially important results.

III. Physiology

A. Epinephrine and norepinephrine content of the reproductive organs: relation to hormonal state and probable functional significance

The epinephrine and norepinephrine content of the female reproductive organs is quite low in comparison with that of other organs such as the heart, spleen, adrenals and some of the male reproductive organs, notably the vas deferens (SJÖBERG, 1967). The absolute values for the content and concentration of epinephrine and norepinephrine in the reproductive organs vary over a wide range not only from species to species, as might be expected, but also within one species when analyzed by different investigators. One possible explanation for these differences is found in the highly individualistic modifications of the basic chemical methods (SHORE and OLIN, 1958; BERTLER et al., 1958; CROUT et al., 1961) used by the different workers for their determinations of tissue catecholamines. Therefore, *changes* in tissue catecholamines under the various experimental situations are probably more significant than the absolute values. For this reason, absolute values will not be given in this review. The interested reader can find these figures in the literature cited in connection with the various topics discussed.

The content of norepinephrine in the internal genitala of most species is considerably higher than that of epinephrine, one exception being the rat (OWMAN et al., 1967). In this species, according to some investigators, the epinephrine content of the uterus (the female reproductive organ usually studied) is higher than in most other organs except the adrenal medulla (RUDZIC and MILLER, 1962a; WURTMAN, CHU and AXELROD, 1963). The norepinephrine content of the rat uterus, however, is lower than that of the other species examined so far (rabbit, cat, human, guinea pig). This finding correlates well with the sparse adrenergic innervation of the rat uterus (NORBERG and FREDRICCSON, 1966). The catecholamines in the rat uterus also possess other unique properties: one, the binding of epinephrine is subject to periodic variations in the normal estrous cycle and during pregnancy, and two, the binding of epinephrine is apparently different from that of norepinephrine.

RUDZIC and MILLER (1962b) were the first to note that the uterine epinephrine stores in the rat were subject to periodic fluctuations during the normal estrous cycle. They found twice as much epinephrine (both concentration and content) in estrus than in diestrus. Similar changes in norepinephrine did not occur. Endogenous epinephrine could also be increased by the administration of diethylstilbesterol. These results were confirmed and extended by WURTMAN, CHU and AXELROD (1963) who, in addition, discovered that the increase in uterine epinephrine concentration during estrus and/or the administration of diethylstilbesterol was related to the ability of the uterus to bind and retain epinephrine within the short interval (about 36 hours) between estrus and diestrus. During this period the capacity of the uterus to bind H^3-epinephrine was increased nearly four-fold, while the binding of H^3-norepinephrine was actually decreased. Since N-methyl-transferase was not found in the rat uterus, it was assumed that the epinephrine must be taken up from the circulation rather than synthesized in the uterus (WURTMAN, AXELROD and KOPIN, 1963). To explore this possibility GREEN and MILLER (1966a) measured the concentrations of plasma epinephrine and norepinephrine in the rat and found epinephrine increased while norepinephrine decreased during estrus. Thus the cyclic variations in uterine epinephrine resulted from variations in the ratio of epinephrine to norepinephrine in the plasma and in the differential ability of the uterus to bind these amines.

During pregnancy, the content of epinephrine in the rat uterus increases about seven-fold, but the concentration of this amine is halved as a result of the large gain in uterine weight (WURTMAN, AXELROD and KOPIN, 1963). Although uterine blood flow increases six-fold during pregnancy the uterine weight increases some thirteen times, hence the fraction of circulating epinephrine delivered to the uterus actually declines. Therefore, the marked fall in uterine epinephrine which accompanies pregnancy in the rat could be due either to a failure of uterine blood flow or the uterine binding capacity for epinephrine to keep pace with uterine growth.

The concentration of uterine epinephrine in post-parturient rats is less than one-quarter that of pregnant and one-half that of nonpregnant uteri. However, the uterus of the post-parturient rat receives and binds about as much circulating epinephrine per unit weight of tissue as the pregnant animal. Therefore, the pronounced decrease in epinephrine concentrations which occurs between six and twelve hours after delivery could be the result of depletion of stored hormone. These interesting findings of WURTMAN, AXELROD and KOPIN (1963) represent the first example of extra-adrenal epinephrine stores which are markedly depleted as a result of a normal physiological event, namely, pregnancy. Similar changes in uterine norepinephrine concentration do not occur.

What is the possible functional significance of these findings? Presumably, norepinephrine in the rat uterus is stored in the nerve terminals and functions as the neurotransmitter. It will be recalled, however, that the histochemical fluorescence methods showed that the distribution of norepinephrine in the rat uterus was confined primarily to the blood vessels and to the utero-tubal junction. Hence, one would anticipate that adrenergic nervous mechanisms do not play a prominent functional role in the myometrium of this species.

MILLER and his colleagues (RUDZIC and MILLER, 1962a and b; SPRATTO and MILLER, 1968a and b) suggested that the endogenous epinephrine may be concerned with the intrinsic regulation of contractility in the rat uterus.

It is well established that epinephrine inhibits the spontaneous activity of the isolated rat uterus, being some 150 times more potent in this respect than norepinephrine (REYNOLDS, 1965). RUDZIK and MILLER (1962a) found that, as the epinephrine content of the rat uterus increased, the rate of contractions of isolated uterine segments decreased. Furthermore, during estrus or after the administration of 17-beta-estradiol (both conditions increase the uterine epinephrine concentration), the spontaneous contractions were significantly lower than in rats in diestrus or proestrus. A careful study has never been made of the frequency of uterine contractions and epinephrine concentration in the pregnant rat, but according to MILLER's hypothesis, the rate of contraction should be very high at term and immediately post-partum. To be physiologically meaningful such a study should be made on the spontaneous motility *in vivo* under conditions where the concentration of uterine epinephrine remained stable during the experimental period. Uterine segments in an isolated organ bath may loose their epinephrine stores over a period of hours. For example, GREEN and MILLER (1966 b) noted that the concentration of H^3-epinephrine in the rat uterus decreased by 15 to 20 % within twenty minutes after removal of the uterine horns from the animal. We can summarize by saying that the physiological significance of the variations in binding of epinephrine in the rat uterus seen during pregnancy and during the estrous cycle remains to be clarified.

Another apparently unique property of the epinephrine in the rat uterus is that its binding and storage sites appear to be different from those for norepinephrine. The findings that (1) epinephrine but not norepinephrine is subject to periodic fluctuations during the estrous cycle and pregnancy; (2) cocaine prevents the uptake of norepinephrine but not epinephrine (WURTMAN et al., 1964; SPRATTO and MILLER, 1968a); and (3) immuno-sympathectomy reduces norepinephrine but not epinephrine content (KLING-MAN, 1965), all suggest that the storage sites for these two amines are different in the rat uterus. Presumably, norepinephrine is stored in the adrenergic neurons. Epinephrine may be stored in some structure (chromaffin cell?) from which it is slowly released, or it might be stored within the smooth muscle cells themselves.

Two questions now arise. First, are these findings in the rat relevant to other species? Second, have they been confirmed by other workers? In a comparative study of the subcellular distribution of epinephrine and nor-epinephrine in the uteri of cats, rats, rabbits and guinea pigs, GUTMAN and WEIL-MALHERBE (1967) could not detect epinephrine in either the particulate fraction, whole homogenate, or soluble fractions of the uterus from any species. These authors state that their method (a modification of the tri-hydroxyindole technique of KAHANE and VESTERGAARD, 1965) was unable to measure total tissue contents of epinephrine below 10 to 20 ng. Similar conclusions have been reached by other workers; for the rat (DIAMOND and BRODY, 1966a; OSKARSSON, 1960; SWEDIN and BRUNDIN, 1968), for the rabbit (MILLER, M. D., and MARSHALL, 1965; SJÖBERG, 1968; ROSENGREN and SJÖBERG, 1968), for the cat (ROSENGREN and SJÖBERG, 1967), for the guinea pig (SJÖBERG, 1968), and for the human (GAFFNEY et al., 1965; OWMAN et al., 1967). On the other hand, BARNEA and SHELESNYAK (1965) detected epinephrine in both the uterus and ovary of the rat. CHA et al. (1965) claim to have found epinephrine in the rabbit and human uterus. It should be mentioned that the values given by MILLER and his colleagues are just at the limit of the sensitivity of his method, around 10 ng total amine content per organ (method of SHORE and OLIN, 1958), and hence changes in epinephrine content are impossible to evaluate statistically. Furthermore, since MILLER and colleagues do not routinely report the percent recovery of known amounts of catecholamines in their samples, it is difficult to assess the reliability of their published figures. On the other hand, the analyses of WURTMAN, AXELROD and KOPIN (1963), most of which were done with tritiated epinephrine, give epinephrine contents in the rat uterus which are well above the limits of their method despite the fact that their per cent recovery was rather low.

Thus aside from the reports for the rat uterus and for the human and rabbit by RUDZIC and MILLER (1962a and b), CHA et al. (1965), BARNEA and SHELESNYAK (1965), significant amounts of epinephrine have never been found in uteri, ovaries or oviducts in any of the species thus far examined. The findings of MILLER and colleagues for the human and rabbit have not been substantiated by other workers.

Changes in norepinephrine content and concentration have been noted in the rabbit uterus and oviduct and in the guinea-pig uterus during pregnancy (SJÖBERG, 1968; ROSENGREN and SJÖBERG, 1968; CHA et al., 1965). In the rabbit, during the first half of pregnancy, there is a marked increase in total norepinephrine content in the uterus, oviduct and vagina. However, the large increase in weight of these organs during pregnancy results in a marked decrease in concentration of norepinephrine in all of these organs except the oviduct. In the oviduct, the increase in norepinephrine content keeps pace with the gain in weight so that the concentration of norepinephrine in this organ is never markedly reduced throughout pregnancy. In the uterus and vagina, a dramatic reduction in norepinephrine content and concentration occurs between 25 and 30 days of gestation (gestation period being 31 days in the rabbit), which brings the tissue concentration of this amine at the end of pregnancy to a level even below that of the non-pregnant uterus. A similar reduction in norepinephrine concentration also occurs in the uterus and vagina of the human and the guinea pig at the end of pregnancy (SJÖBERG, 1967).

The functional significance of these conspicuous alterations in norepinephrine during pregnancy must await an elucidation of the overall role of adrenergic nervous mechanisms in the female reproductive processes. Nevertheless, the dramatic reduction in the concentration of adrenergic neurotransmitter (norepinephrine) and in the effective density of adrenergic nerves within the myometrium and vagina (see p. 20) indicates that the amount of transmitter released per unit area of muscle is greatly diminished near the end of pregnancy. As a result, the efficacy of the neurogenic control over myometrial activity would be minimal at this time. Immediately before parturition there is an increase in spontaneous, myogenic motility in the uterus and in the myometrial sensitivity to oxytocin (REYNOLDS, 1965). Both of these changes would assure strong, coordinated contractions, perhaps reducing the necessity for a direct neural mediation.

Estrogen treatment for 14 consecutive days produces changes in norepinephrine in the uterus and vagina of rabbits similar to those observed during the first half of pregnancy, i.e. a pronounced increase in content but reduction in concentration (ROSENGREN and SJÖBERG, 1967). These findings by the Swedish workers are in contrast to the earlier ones of MILLER and MARSHALL (1965) who could not detect a significant change either in uterine norepinephrine content or concentration after 5 days of estrogen treatment. The discrepancy between the results of these two groups probably resides in the duration of the estrogen treatment, a period longer than 5 days might be needed to induce the change in norepinephrine content.

In addition to these changes observed during pregnancy and after estrogen administration, there are also differences in norepinephrine content and concentration among the various reproductive organs of some species. For example, the ovaries of the cat and of the human contain a higher concentration of norepinephrine than any other portion of the reproductive tract in these species. In fact, the ovary of the cat has the highest concentration of norepinephrine of any of the female genitalia in all species thus far examined (rabbit, guinea pig, human, monkey, rat). All of the female genitalia in the cat have significantly higher concentrations of norepinephrine than those of the other species. The human oviduct, particularly the isthmus, contains a higher concentration of norepinephrine than the uterine corpus or fundus. The norepinephrine concentration of the cervix is, however, about equal to that of the oviduct (OWMAN et al., 1967).

Removal of the inferior mesenteric ganglia and both hypogastric nerves is practically without effect on the norepinephrine content or concentration in the female reproductive tract of the cat, guinea pig, or rat, the only species studied so far (OSKARSSON, 1960; ROSENGREN and SJÖBERG, 1967; SJÖBERG, 1967). This is additional evidence that the adrenergic nerve fibers running to these organs arise mainly from peripheral ganglion cells.

B. Adrenergic nervous influences

1. Some electrophysiological properties of autonomic nerves (hypogastric and uterine nerves)

The hypogastric nerve is the one usually selected for studies on the effects of extrinsic nerve stimulation on the reproductive organs. This nerve contains myelinated and unmyelinated, pre- and postganglionic fibers including efferents to the smooth muscles of the blood vessels and of the reproductive organs and afferents from sensory receptors in these organs (LANGLEY and ANDERSON, 1896; SCHOFIELD, 1952; REYNOLDS, 1965; BOWER, 1966a and b; FERRY, 1967). Some of the electrophysiological characteristics of various types of mammalian nerve fibers are summarized in Table 1. According to this table, the fiber spectrum of the hypogastric nerve includes types A (fast, sensory); B (preganglionic); and C (postganglionic and sensory).

Table 1. *Characteristics of major types of mammalian nerve fibers.* (Modified from GRUND-FEST, H.: Bioelectric potentials, Ann. Rev. Physiol. 2, 213, 1940; from DAVIES, P. W., 1968)

	Type A motor, fast sensory myelinated	Type B autonomic, preganglionic, myelinated	Type C autonomic, post- ganglionic, sensory, unmyelinated
Fiber diameter (µm)	20 to 1	< 3	0.5 to 1.0
Conduction velocity (m/sec)	100 to < 5	$14 < 3$	< 2
Action potential duration (msec)	0.4 to 0.5	1.2	2.0
Absolute refractory period (msec)	0.4 to 0.5	1.2	2.0

In comparison with the large somatic motor axons, autonomic nerve fibers have relatively small diameters and therefore slower conduction velocities. The duration of the conducted action potential and the refractory period are longer in autonomic axons than in the somatic motor nerves and hence the upper limit of the rate of impulse discharge in the autonomic efferents is considerably below that for the somatic motor system. From studies on the frequency of action potential discharge in various autonomic efferents, including fibers in the hypogastric nerve, it has been concluded that the normal spontaneous discharge rate is usually low, between 1 and 4 impulses per second and rarely exceeds 10 to 20 per second (ADRIAN et al., 1932; BRONK et al., 1938; CELANDER, 1954; BOWER, 1966a).

FOLKOW (1952) concluded that the normal peripheral resistance of the small blood vessels is maintained by a discharge frequency in the vasomotor nerves as low as 1 to 2 impulses per second. The increase in frequency accompanying an intense stimulation of the vasomotor center never exceeds 6 to 8 per second. On the basis of these findings, CELANDER (1954) concludes that the normal range of impulse discharge in the autonomic nervous system is limited and probably never exceeds 10 impulses per second even during intense reflex activation. This fact should be kept in mind in any investigation of the effects of adrenergic nerve stimulation on the female reproductive organs. The frequency of stimulation which would be expected to produce a physiological response of the effector organ would most probably reside between 1 and 10 pulses per second.

A detailed account of the electrophysiology of uterine nerves has recently been published by BOWER (1966a and b), who succeeded in recording action potentials from these nerves in the intact rabbit. In this animal the uterine nerves are about 50 to 200 μm in diameter. They arise from ganglion cells within the pelvic plexus and pass into the uterine horns along the broad ligament (LANGLEY and ANDERSON, 1896). When BOWER (1966a) stimulated the hypogastric nerve a compound action potential appeared in the uterine nerves but stimulation of the pelvic nerves was without an effect. Hence, in the rabbit, pelvic nerves (cholinergic) do not relay to the uterus via the uterine nerves. This is the first direct electrophysiological evidence in support of the report of LANGLEY and ANDERSON (1896) that stimulation of the pelvic nerves in the rabbit had no effect on uterine motility.

Two types of spontaneous action potentials were recorded from the uncut uterine nerves; one travelled at about 4 m/sec and probably came from type A sensory afferents from receptors in the broad ligament, the other travelled at velocities from 0.4 to 1.4 m/sec and probably represented C fibers, including sensory afferents and postganglionic efferents. A continuous, spontaneous discharge of action potentials was recorded from the central ends of cut uterine nerves. The frequency of this discharge was remarkably regular between 1 and 4 impulses per second (BOWER, 1966b). Although this resting discharge was more intense late in pregnancy, there was no correlation between the rate of discharge and uterine contractions in estrus, diestrus or pseudo pregnant animals. The finding that the frequency of discharge was increased during asphyxia and reduced by stimulation of the central end of the depressor nerve, led BOWER (1966b) to suggest that it may represent tonic vasomotor activity rather than anything directly related to the myometrium.

Evidence that the majority of the efferent fibers in the uterine nerves are postganglionic comes from BOWER'S (1966a) experiments with ganglion-blocking agents. These agents substantially reduced the electrical activity in the uterine nerves during stimulation of the hypogastric nerve. Hence this is the first *direct* electrophysiological evidence for the presence of peripheral ganglion cells in the autonomic pathways to the female reproductive organs. Even in the presence of ganglion blockers, however, a few uterine nerve fibers still fired in response to hypogastric stimulation, indicating that a few efferent fibers to the myometrium did not synapse in the pelvic ganglia. This might explain why previous workers (Schofield, 1952; Varagic, 1956; Miller and Marshall, 1965) could not abolish uterine contractions in response to hypo-gastric stimulation even with huge doses of hexamethonium. These workers used estrogen-dominated rabbits, and it is well established that the myo-metrium in such animals behaves like a "physiological syncytium", i.e. excitation of one part of the muscle spreads rapidly throughout the tissue via cell-to-cell, myogenic conduction. As a result, the uterus would be capable of giving a full-sized contraction in response to a localized stimulation from only a few adrenergic nerve terminals since neurogenically-induced excitation could spread over the muscle via the myogenic conduction process.

Another important finding of Bower was the fact that stimulation of the hypogastric nerve with single shocks often resulted in a prolonged after discharge in the uterine nerve fibers. The after discharge was abolished when the uterine nerves were cut between the pelvic ganglia and the uterus, indi-cating that the origin of the repetitive response resided in the ganglion cells. Thus the pelvic ganglia may be quite different from other autonomic ganglia (stellate, superior cervical) where the input output relations are usually one-to-one (Bronk et al., 1938). The finding of repetitive discharge in the uterine nerves raises the possibility of a prolonged release of transmitter at the post-ganglionic nerve terminals in response to a single preganglionic impulse. Thus if the innervation density is such that not all the myometrial cells receive discrete nerve axons (as seems to be the case in some species) (Silva, 1966), a prolonged release of transmitter would be advantageous when many of the effector cells need to be activated by generalized diffusion of transmitter. Moreover, a repetitive discharge in the postganglionic neurons would favor a long-lasting effect of preganglionic stimulation.

2. Effects of adrenergic nerve stimulation
a) Uterus

The most extensive studies on the effects of peripheral nerve stimulation on the female reproductive organs are those involving the hypogastric nerve and the uterus. A summary of some of the findings appears in Table 2, and shows that the response to nerve stimulation depends upon species, phase of the estrous cycle, and the presence or absence of pregnancy.

The classical studies of the effects of nerve stimulation were those of Langley and Anderson (1895a) who applied both anatomical and physio-logical techniques to the problem. Although their methods of nerve stimulation were of necessity crude—an induction coil stimulator with the stimulus intensity monitored by the investigator's tongue and changes in uterine motility often observed only visually—nevertheless, the published reports of their work are models for accuracy of observation and clarity of expression. Hypogastric stimulation elicited a contraction of the uterus of the rabbit and the cat, although in the latter species the effect was less consistent. Blanching and pallor of the uterus usually accompanied the contractions, indicating a vasoconstrictor component to the nervous influence. Since atropine did not prevent the response, while epinephrine mimicked it, LANGLEY and ANDERSON concluded that the hypogastric nerve was predominantly adrenergic. As mentioned earlier (p. 17), their experiments with nicotine led to the idea that the majority of the adrenergic ganglia lay close to the uterus.

Shortly after these studies, CUSHNY (1906) drew attention to the remarkable fact, noted almost simultaneously by DALE (1906), that stimulation of the hypogastric nerves in the cat caused powerful contractions of the uterus in the pregnant cat at mid-term while the uterus of the non-pregnant or virgin cat always relaxed. Near the end of pregnancy the effect of nerve stimulation became diphasic, an initial contraction followed by a relaxation. DALE coined the term "pregnancy reversal" for this effect and found it also was mimicked by epinephrine. He further showed that the motor effects of nerve stimulation could be prevented by extracts of ergot. The phenomenon of "pregnancy reversal" has been repeatedly confirmed and extended to other species, notably the rabbit (RUDOLPH and IVY, 1930; SAUER et al., 1935; GUSTAVSON and VAN DYKE, 1931; MANN and WEST, 1951; VOGT, 1965).

Naturally, a host of speculations was put forward to explain "pregnancy reversal", the most prominent being that the hypogastric nerves contained both excitatory and inhibitory fibers and during pregnancy there was a change in the relative dominance of the two systems. Emphasis was generally placed on changes in the nerves rather than in the effector cells (see GRUBER, 1933, for erroneous speculations). However, LANGLEY (1901 b) with his usual perspicacity suggested that the effector cells might contain both excitatory and inhibitory "substances" and that the response of the myometrial cell then would depend upon the relative proportion of these two substances released during nerve stimulation. GADDUM (1926), studying the effects of epinephrine after pretreatment of an isolated rabbit uterus with ergotamine, advanced a theory that epinephrine (at that time believed to be synonymous with adrenergic nerve stimulation) acted on a specific area of the myometrial cell, and that ergotamine blocked a fraction of this area so that the concentration of epinephrine had to be increased to overcome the block. Thus, GADDUM's ideas focused on the effector cell rather than on the nerve fibers. It is interesting that, even in 1933, GRUBER still favored the theory that some change takes place within the nerves themselves during pregnancy. He dismisses the whole idea of chemical transmission of nerve impulses and the possible effects of the neurotransmitter on effector cells, with one brief statement at the very end of his review. This despite the fact that LOEWI had demonstrated chemical transmission in 1921 and CANNON and associates were publishing their important work on adrenergic transmission in the early 1930's!

Table 2. *Effects of hypogastric nerve stimulation on uterine motility*

Animal	Stimulus parameters	Conditions (hormonal)	Results	Reference
Rabbit				
in situ	induction shocks	not stated	Contraction of uterus; pallor of uterus (mimicked by epinephrine)	LANGLEY and ANDERSON (1895b)
in situ	induction shocks	pregnant	Contraction followed by reduction of spontaneous contractions	CUSHNY (1906)
in situ	induction shocks	non-pregnant	Contraction followed by relaxation; prevented by ergot extract	DALE (1906)
in situ	induction shocks	non-pregnant / pregnant	Contraction and blanching / Less sensitive to nerve stimulation	RUDOLPH and IVY (1930)
in situ	induction shocks	castrate / castrate and estrogen / pseudopregnant	Small contraction / Strong contraction mimicked by epinephrine / No effect or relaxation	SAUER et al. (1935)
in situ	induction shocks	castrate / non-pregnant / pregnant	Small contraction / Contraction / No response	LABATE (1941)
in situ	square wave (1 msec) 50 p/sec [a]	castrate and estrogen / term pregnant / post partum	Contraction / No effect or relaxation / No effect or relaxation	SCHOFIELD (1952)
in vitro	square wave (1 msec) 10—50 p/sec	virgin, untreated	Contraction	VARAGIĆ (1956)
in vitro / in situ	square wave (1 msec) 5—50 p/sec	castrate and estrogen	Contraction / Blocked by phentolamine / Blocked by dihydroergotamine	SETEKLIEV (1964)

in vitro	immature immature and estrogen immature and estrogen and progesterone	square wave (1 msec) 5—50 p/sec	Contraction (feeble) Contraction (strong) (blocked by phentolamine) Relaxation (blocked by propranolol)	MILLER and MARSHALL (1965)
Cat in situ	non-pregnant	induction shocks	Inconsistent response	LANGLEY and ANDERSON (1895b)
in situ	non-pregnant pregnant	induction shocks	Relaxation Contraction (mimicked by epinephrine)	CUSHNY (1906)
in situ	non-pregnant pregnant	induction shocks	Relaxation Contraction (prevented by ergot extract) (mimicked by epinephrine)	DALE (1906)
in situ	non-pregnant pregnant	induction shocks	Relaxation Contraction (mimicked by norepinephrine and epinephrine)	MANN and WEST (1951)
in situ	non-pregnant pregnant	square wave (1 msec) 10—30 p/sec	Relaxation (reversed by DCI and pronethalol) Contraction (mimicked by epinephrine and norepinephrine)	VOGT (1965)
Rat in situ	castrate non-pregnant pregnant (mid-term)	induction shocks	No effect Relaxation Contraction, or slight relaxation	LABATE (1941)
Dog in situ	non-pregnant	induction shocks	Contraction followed by inhibition	CUSHNY (1906)
in situ	post partum	induction shocks	No effect	RUDOLPH and IVY (1930)

3*

Table 2. (continued)

Animal	Stimulus parameters	Conditions (hormonal)	Results	References
Guinea pig				
in situ	induction shocks	non-pregnant	Contraction (abolished by ergot extract)	DALE and LAIDLAW (1912)
in situ	induction shocks	non-pregnant pregnant	Relaxation Relaxation	GUNN and FRANKLIN (1922)
in vitro	square wave[b] (1—2 msec) 5—10 p/sec	diestrus estrus	No effect or relaxation Contraction	PENNEFATHER and ISAAC (1967)
in situ	square wave (1 msec) 20—30 p/sec	estrus	Contraction	RÜSSE and MARSHALL (1969)
Human				
in situ	condensor discharge 4 p/sec	follicular phase luteal phase	Small contraction followed by relaxation Strong contraction followed by relaxation	CALDEYRO-BARCIA and ALVAREZ (1954)
in situ	square wave (1 msec) 5 p/sec	pregnant (early)	Strong contraction	THEOBALD (1968)
Monkey				
in situ	induction shocks	non-pregnant	Small contraction	REYNOLDS (1965)
in situ	induction shocks	non-pregnant	Contraction and vasoconstriction	LABATE (1941)

[a] pulses/sec.
[b] Ovarian nerves.

We now know that stimulation of the hypogastric nerve in the cat releases both epinephrine and norepinephrine irrespective of the hormonal or gravid state of the animal (MANN and WEST, 1951; VOGT, 1965). The response of the muscle to nerve stimulation, therefore, resides in the effector cell. Whether the response is excitatory or inhibitory is determined by the reaction of the neurotransmitter with one or more specific molecular sites on or within the effector cell, in this instance the myometrial cell. These sites are called "adrenoceptive sites" (DALE, 1954), "adrenoreceptors" (BOWMAN and NOTT, 1969), or "*adrenoceptors*"[1]. We are as yet ignorant of the precise chemical and morphological nature of these reactive regions. The familiar classification of adrenoceptors into alpha and beta receptors (AHLQUIST, 1948) is now generally accepted. All uteri thus far examined contain both alpha (excitatory) and beta (inhibitory) receptors (MILLER, 1967). Norepinephrine has a greater affinity for alpha receptors while epinephrine has about equal affinity for both alpha and beta receptors (AHLQUIST, 1948, 1966; MILLER, 1967). The effect of adrenergic nerve stimulation depends, among other things, on the relative dominance of one or the other receptor. It will be excitatory if the alpha receptors predominate, and inhibitory if the beta receptors are in the majority. The contraction of the pregnant cat uterus is prevented (and often converted into a relaxation) by the alpha receptor-blocking agent, phentolamine, and the inhibition of the virgin uterus by the beta receptor-blocking agent, propranolol (VOGT, 1965).

The relative dominance of either alpha or beta receptors is apparently under hormonal control in the cat uterus. The implication of the ovarian hormones in "pregnancy reversal" was first suggested by GUSTAVSON and VAN DYKE (1931). They found that the uterus of a spayed cat injected with extracts of the corpus luteum contracted in response to hypogastric nerve stimulation while that of a cat injected with follicular fluid would relax. Unfortunately these experiments have never been repeated using purified preparations of progesterone and estrogen (which were not available in 1931).

In the rabbit "pregnancy reversal", although less pronounced, is just the opposite of that in the cat, i.e. the non-pregnant uterus of the rabbit contracts strongly when the hypogastric nerve is stimulated, while the uterus of the pregnant animal either relaxes or does not respond (Table 2). Whether the

---

1 The term "*adrenoceptor*" is introduced as a synonym of "adrenoceptive site" (DALE, 1954) to take the place of the inaccurate expression "adrenergic receptor". (In the same way "cholinoceptor" should be used instead of "cholinergic receptor".) This is in agreement with DALE's admonition that, with the loss of their precision, scientific terms loose their value, and that, therefore, "adrenergic" (or "cholinergic") should be used, as originally conceived, only in reference to neurons, nerve fibers and their effects (DALE, 1933, 1953; FELDBERG, 1969). The terms "alpha receptor" and "beta receptor" require no further characterization because they refer exclusively to adrenoceptors.

hypogastric nerve in the rabbit releases both epinephrine and norepinephrine upon stimulation has never been determined. Pretreatment of immature rabbits with estrogen causes the uterus to contract on hypogastric nerve stimulation, an effect which is prevented by an alpha receptor-blocking agent. Pretreatment with estrogen followed by progesterone results in an inhibitory response to nerve stimulation. Inhibition can be prevented by beta receptor-blocking agents (SETEKLIEV, 1964; MILLER and MARSHALL, 1965).

Thus estrogen administration appears to promote beta receptor domination in the uterus of the cat while in the rabbit it favors alpha receptor domination. Just the opposite is true for progesterone. This hormone favors alpha receptor dominance in the cat and beta receptor dominance in the rabbit. Whether the changes seen in pregnancy can be explained as simply the result of the relative dominance of estrogen or progesterone seems highly unlikely in view of the many other complex hormonal changes taking place during gestation.

The effects of adrenergic nerve stimulation on uterine motility has been studied more extensively in the rabbit and in the cat than in most other species (Table 2). CALDEYRO-BARCIA and ALVAREZ (1954) stimulated the hypogastric plexus in women during surgery for relief of dysmenorrhea and thereby caused the whole uterus to contract. The contractions were stronger during the luteal than during the follicular phase of the cycle, implying a possible hormonal regulation of the response. Strong contractions of the human uterus during early pregnancy were also noted by THEOBALD (1968) during stimulation of the hypogastric plexus.

A careful and systematic analysis of the effects of adrenergic nerve stimulation on a variety of species during pregnancy and at various times in the reproductive cycle is needed before any functional significance can be assigned to the phenomenon of "pregnancy reversal". The phenomenon, however, is of considerable interest in its own right since it is one of the few, if not the only, example of a definite physiological (more precisely endocrinological) regulation of adrenoceptors in autonomic effector cells.

Unfortunately, in many of the experiments involving the effects of the hypogastric nerve stimulation on the uterus, the frequency and intensity of nerve stimulation are well above that believed to be in the physiological range. Many of the earlier workers were forced by necessity to use induction coil stimulators since the modern pulse generators had not yet been developed. Hence the frequency (faradic) and intensity of stimulation could not be controlled and was undoubtedly well beyond the physiological range. Even when modern square-wave stimulators are used, most of the experiments are done with a frequency of stimulation that gives the optimum response of the effector organ. This is usually between 20 and 50 pulses per second. A recent study (RÜSSE and MARSHALL, 1969) on the non-pregnant guinea pig *in vivo* showed that, when the hypogastric nerve was stimulated at frequencies between 1 and 5 pulses per second and at an intensity of 1 mA, there was no observable effect on the uterus. This frequency is well within the physiological discharge rate of autonomic fibers (CELANDER, 1954). Only when the stimulation was increased to between 10 and 30 pulses per second would the uterus contract. However, at the lower frequencies (1 to 5 pulses per second) even though there was no observable direct action on the uterus its response to oxytocin was markedly potentiated. These findings prompted the suggestion that one possible function of adrenergic nerve activity may be to alter the uterine sensitivity to circulating hormones. This effect is observed experimentally only if the nerves are stimulated at rates within the physiological range. In this connection it is interesting that, in most of the studies shown in Table 2, the direct effects of nerve stimulation on the motility patterns of the uterus became obvious only at stimulation frequencies above 5 pulses per second, and generally were most obvious between 20 and 30 pulses per second.

Virtually no information is available concerning the electrophysiological consequences of stimulation of adrenergic nerves on the uterus or on any other organs in the female reproductive tract. Several investigations of the effects of hypogastric nerve stimulation on the gross electrical activity of the rabbit uterus were reported many years ago (HASAMA, 1931; MORISON, 1940) before the more refined techniques of microelectrode or suction electrode recording were developed. Nevertheless, these early studies, utilizing non-polarizable extracellular electrodes applied to the uterus in situ, did show that during stimulation of the hypogastric nerves in the virgin rabbit there was an overall depolarization of the uterus, while the pregnant uterus either showed no change or a slight reduction in spontaneous electrical waves. Since uterine contractions were not registered, it is difficult to tell whether these electrical changes were real or were merely the result of movement artifacts.

Studies similar to those of BURNSTOCK and HOLMAN (1961) on the electrophysiology of neuro-effector transmission in the vas deferens at the cellular level have never been done on the myometrium, or on any other smooth muscle in the female reproductive tract. It is likely, however, that the basic aspects of the transmission processes are common to all autonomic neuro-effector junctions (BURNSTOCK and HOLMAN, 1966a). It might be appropriate, therefore, to consider some of these aspects and how they might theoretically apply to the neuro-effector transmission in the myometrium. Hopefully, by the time the next review of the female genitalia is made, experiments similar to those on the vas deferens will have been performed on the myometrium.

When the nerve impulse invades the adrenergic nerve terminal it releases some of the norepinephrine (and epinephrine if present) stored there. The neuro-transmitter then diffuses across the intervening space between the nerve and effector cell and reacts with a part of the effector cell membrane which LANGLEY (1901a) called the "receptive substance". Today it is called a receptor, in this case, the adrenoceptor. The reaction of the transmitter with the receptor results in the response of the effector cell. The morphological sites for transmitter release are the beaded varicosities along the small nerve fibers visualized by fluorescence microscopy and the regions of the axon containing the granular vesicles seen in the electron microscope. As mentioned previously, nerve axons and the smooth muscle cell membranes are not specialized at their points of close proximity. In autonomic neuro-effector systems, the neuron appears to release transmitter at various regions all along its terminal axon rather than at a single, discrete nerve ending as at the somatic neuro-muscular junction (BURNSTOCK and HOLMAN, 1966a). Some of these regions of release lie in close proximity to the effector cells while others lie at greater distances in the autonomic ground plexus.

The vas deferens is so densely innervated that each smooth muscle cell lies near or actually next to a nerve axon, and for this reason it was selected by BURNSTOCK and HOLMAN (1961) for their studies on neuro-effector transmission in the autonomic nervous system (HOLMAN, 1967). These investigators introduced microelectrodes into single smooth muscle cells in the vas deferens of the guinea pig and observed the changes in transmembrane potentials produced by stimulation of the hypogastric nerve. Nerve stimulation caused a transient, localized depolarization of the muscle membrane (excitatory junction potential) in about every cell impaled with the electrode. With repeated stimulation the junction potentials summed and when threshold was reached an action potential appeared and the muscle contracted. In the absence of nerve stimulation a random discharge of small, spontaneous depolarizations (spontaneous excitatory junction potentials) was noted. The spontaneous potentials probably represent the release of "packets" of neurotransmitter (HOLMAN, 1967). Similar findings for excitatory adrenergic transmission have been reported for the retractor penis of the dog (ORLOV, 1961) and the smooth muscle of arterioles (SPEDEN, 1964), and suggest that the mechanism for release of norepinephrine from adrenergic nerves is much like that for acetylcholine at the neuromuscular junction in skeletal muscle. Inhibitory junction potentials characterized by a hyperpolarization of the cell membrane in response to adrenergic nerve stimulation in the intestine have also been noted (BURNSTOCK, CAMPBELL et al., 1963).

Although similar studies have never been reported for the effects of adrenergic nerve stimulation on the female reproductive organs, it seems reasonable to assume that the transmission processes here are basically similar to those just described for other autonomic effectors. However, the female reproductive organs are not as densely innervated as the vas deferens (SJÖBERG, 1967), and the number of nerve axons in close proximity to the smooth muscle cells is considerably less in the uterus and in the oviduct than in the vas deferens (cf. SILVA, 1966; KUSHIYA, 1968). In the rat uterus and rabbit oviduct many of the nerve axons were observed to lie at some distance from the muscle cells. Hence transmitter would reach most of these muscle cells by diffusion. However, the uterus and oviduct, unlike the vas deferens, are often spontaneously active, indicating the presence of myogenic pacemakers and conduction via electrical coupling between muscle cells (BURNSTOCK et al., 1963). It is conceivable that, in muscles such as those of the uterus or oviduct which contract (or relax) in response to nerve stimulation, a few cells would be influenced directly by transmitter, some cells would be activated by diffusion of transmitter, other cells would be activated by electrotonic spread of depolarization (or hyperpolarization) caused by transmitter release on neighboring cells, but most cells would be activiated by action potentials myogenically conducted from "target" cells receiving direct innervation. Therefore, junction potentials similar to those observed in the vas deferens would be recorded only if the microelectrode happened to impale a cell whose membrane was in close apposition to a nerve axon. The animal of choice for such studies would be the female cat whose reproductive organs have the highest catecholamine content and innervation density of all species studied thus far (SJÖBERG, 1967). Until such experiments have been done, further speculations about the cellular physiology of neuroeffector transmission in the female internal genitalia are unwarranted.

b) Cervix and vagina

Most workers state that, in the cat and rabbit, the reactions of the cervix and vagina are identical to those of the uterus during hypogastric nerve stimulation (LANGLEY and ANDERSON, 1895a; DALE, 1906; RUDOLPH and IVY, 1930; SCHOFIELD, 1952). GUNN and FRANKLIN (1922) took exception to this because they found that, in the cat, hypogastric stimulation always contracted the vagina even when the uterus was relaxed (e.g. in the virgin cat). On the basis of this observation, they suggested that the simultaneous contraction of the vagina and relaxation of the uterus elicited by adrenergic nervous activity during coitus might aid in sperm transport to the uterus. Apparently, these interesting experiments have never been repeated.

c) Oviduct

As noted earlier, the oviduct in both the human and rabbit is abundantly supplied with adrenergic nerves from the hypogastric plexus, the circular muscle of the isthmus being especially well endowed (BRUNDIN, 1964, 1965). Stimulation of the hypogastric nerve (at about 17 pulses/sec) causes a contraction of the rabbit oviduct *in situ* as indicated by an increase in intraluminal pressure. Stimulation (30 to 50 pulses/sec) of the perivascular nerves around the middle tubal branch of the ovarian artery running to the isolated human fallopian tube causes a contraction of that organ followed by an increase in its spontaneous motility (NAKANISHI et al., 1967). The response is unaffected by ganglion-blocking agents but it is converted to an inhibition in the presence of an alpha receptor blocker (phenoxybenzamine). The inhibition is abolished by propranolol (beta receptor blocker). Subsequently, NAKANISHI and WOOD (1968) found that the isthmus of the human oviduct is more sensitive to nerve stimulation than the ampulla, and suggested that the isthmus might act as a sphincter. Many years ago BURDICK and PINCUS (1935) showed that, following ovulation in the mouse and in the rabbit, ova were retained in the ampulla of the oviduct for a comparatively long time before continuing down to the uterus. This observation prompted them to postulate that a "tubal locking mechanism" may be responsible for retention of ova in the ampulla. A sphincter at the isthmic end of the oviduct near the utero-tubal junction whose activity is controlled by adrenergic nerves could be responsible for this locking mechanism which prevents the ova from reaching the uterus (BRUNDIN, 1964). Functional blockage as a result of contraction of the lumen of the oviduct during adrenergic nerve activity in individuals undergoing stress or fear, may thus affect transport of ova and fertility. It would be of interest to know whether utero-tubal spasm in humans (as shown by insufflation techniques) is relieved by alpha receptor-blocking agents as suggested by the results of NAKANISHI and WOOD (1968).

No studies have been reported to date on the influence of the ovarian hormones or of various reproductive states on the tubal response to adrenergic nerve stimulation.

d) Uterine blood vessels

The hypogastric and uterine nerves contain vasomotor fibers which when stimulated invariably cause contraction of the small arteries supplying the reproductive organs (GREISS, 1963). These vasomotor effects have been known since the work of LANGLEY and ANDERSON (1895 a and b, 1896) who noted "pallor" of the uterus during hypogastric stimulation. The effects, therefore, may complicate experiments concerned with nervous influences on the motility of the uterus or other organs in the intact animal. This is particularly true if the periods of nerve stimulation are continued for many minutes. VOGT (1965), who had to stimulate the hypogastric nerves in the cat for periods up to 10 minutes in order to collect sufficient venous effluent to measure the neurotransmitter content, mentions the depression of uterine motility that resulted from the intense vasoconstriction. The uterine quiescence following prolonged hypogastric stimulation in the rabbit was ascribed to vasoconstriction and subsequent reduction in uterine blood flow (SCHOFIELD, 1952).

GREISS and GOBBLE (1967) stimulated the hypogastric plexus (15 pulses per second for one minute) in pregnant anesthetized ewes while measuring uterine blood flow with implanted flowmeters, and found intense vasoconstriction in both uterine horns. In a series of experiments on unanesthetized pregnant ewes with flowmeters implanted in their uterine arteries, GREISS et al. (1967) found that the resting tone in the adrenergic vasomotor fibers was minimal and, as a result, the vascular bed was fully dilated and functioning near maximal efficiency as would be predicted during pregnancy (REYNOLDS, 1965). However, if the animal was startled or subjected to a stressful stimulus, uterine blood flow was reduced by at least 25 per cent. All of these experiments emphasize the fact that stimulation of the adrenergic nerves to the uterus in the intact animal usually causes a profound change in uterine blood flow. Therefore, in all *in vivo* experiments involving adrenergic nerve stimulation to the reproductive organs, care must be taken to separate the indirect effects of changes in blood flow from the direct actions of the neurotransmitter on the effector organ. This can best be done by simultaneous monitoring of uterine blood flow and restricting the nerve stimulation to short periods and to frequencies below 8 to 10 pulses per second (FOLKOW, 1952).

In the latter connection, a recent series of experiments by BELL (1968) on an isolated segment of uterine artery from the guinea pig indicated that, when the periarterial nerves were stimulated at frequencies between 10 and 70 pulses per second, a pronounced constriction of the vessel occurred. This effect was prevented by alpha receptor-blocking agents and mimicked by norepinephrine. However, at frequencies of stimulation between 1 and 5 pulses per second no effect was noted.

3. Afferent input from pelvic organs to central nervous system; reflex regulation of uterine activity

The hypogastric and uterine nerves contain, in addition to the vasomotor and myomotor efferent fibers, sensory afferents some of which probably come from mechanoreceptors in the uterus, cervix, vagina and broad ligament (BOWER, 1966b). Evidence for a central projection of these sensory afferents comes from the experiments of BARRACLOUGH and CROSS (1963) who recorded unitary electrical activity from areas in the hypothalamus and diencephalon in female rats. Units in the hypothalamus had a spontaneous firing rate of from 1 to 10 pulses per second. The firing rate of these units could be accelerated by mechanical probing of the cervix and by pain or cold stimuli. The level of responsiveness was highest in diestrus and lowest in estrus, and in all animals progesterone, i.v., produced a selective depression of the response of the neurons to cervical probing. The progesterone effect lasted for about one hour.

Potentials were also evoked in the hypothalamus and in the somatosensory area of the cerebral cortex of cats when the uterus was stimulated electrically (ABRAHAMS et al., 1964).

These modern experiments support the earlier findings of FERGUSON (1941) who noted that distension of the one uterine horn, the cervix, or the vagina in rabbits enhanced the motility in the other horn. This reflex was abolished by section of the spinal cord at midthoracic level or by cauterization of the pituitary stalk. The uterine contractions were believed to result from reflex release of oxytocin. Later, Cross (1958) tried to repeat these experiments but could only confirm the reflex stimulation of the uterus after distension of the vagina. Furthermore, the reflex seemed to have two components, one of which was similar to the prolonged increase in uterine motility elicited by oxytocin and was abolished by removal of the pituitary gland, and the other was a strong, transient contraction resembling that of injected norepinephrine. The latter response was not affected by removal of the pituitary gland, and was most certainly mediated via efferent adrenergic nerves. Therefore, it is tempting to speculate that the relay between the uterus, vagina, cervix and the hypothalamus is concerned not only with the release of oxytocin, but also with the reflex activation of peripheral adrenergic neurons to the uterus. Adrenergic nerve stimulation may not only augment contractions of the uterus directly, but it may also increase the sensitivity of uterus to oxytocin (cf. Rüsse and Marshall, 1969).

4. Central adrenergic mechanisms

Although the main emphasis of this review is on peripheral adrenergic mechanisms, it is, of course, well established that the central nervous system influences reproductive functions. The most generally recognized pathway for this action is via the hypothalamus to the pituitary gland. Much attention has been directed to the mechanisms of hypothalamic control of the release of adenohypophysial gonadotrophins during the development and maintenance of sexual function, including ovulation and lactation (Everett, 1964). Recent studies indicate that adrenergic pathways may play a significant role in controlling the ovulatory surge of luteinizing hormone (LH) in the rat (Gaunt et al., 1963; Coppola et al., 1966; Hopkins and Pincus, 1966; Meyerson and Sawyer, 1968). Indirect evidence for the existence of such pathways comes from the findings of Hopkins and Pincus (1966) who noted that, in the rat, reserpine exerted a profound inhibition on the ovulatory response to human chorionic gonadotrophin (HCG). This action was not related to a direct antagonism between reserpine and HCG or to the destruction of ova by reserpine since the ovaries of the treated animals contained follicles but were devoid of corpora lutea. Therefore, the action of reserpine was thought to involve LH synthesis and/or release. Peripheral depletors of catecholamines, such as tyramine and guanethidine, had no effect on the ovulation, but the ovulatory blockade induced by reserpine could be reversed by monoamine oxidase inhibitors (Coppola et al., 1966). Thus reserpine must have affected a central adrenergic pathway.

It is well established that an ovulatory surge of LH, triggered by central nervous activation of the adenohypophysis, invariably occurs in rats between 1,400 and 1,600 hours on the day of proestrus (SAWYER, 1963). With this in mind, MEYERSON and SAWYER (1968) injected reserpine into rats on the day of proestrus just prior to the critical period for the central nervous activation of LH release. In all cases ovulation was prevented, but in no case was it prevented if reserpine was injected after the critical period for LH release. These ingenious experiments also placed the effect of reserpine within the mechanism of LH release rather than directly on the ovary. An effect solely on the pituitary seems unlikely in view of the evidence that the adenohypophysis lacks a nerve supply concerned with the output of LH and, more important, its monoamine content is not depleted by reserpine (DAHLSTRÖM and FUXE, 1966). The hypothesis put forward by MEYERSON and SAWYER (1968) suggests that adrenergic nervous activity, originating in the hypothalamus, normally acts to stimulate LH release and to inhibit synthesis or release of luteotrophic hormone (LtH). When this adrenergic influence is diminished or absent, then LH secretion is decreased and LtH is increased. The main focus is on the hypothalamus since it is the "final common pathway" between the central nervous system and the pituitary. Furthermore, it is very rich in monoamines (CARLSSON et al., 1962). Morphologic evidence for the existence of adrenergic pathways between the hypothalamus and hypophysis has been shown by histochemical fluorescence techniques, in cats, rats, rabbits, mice, hamsters and guinea pigs (CARLSSON et al., 1962). Fine, varicose, catecholamine-containing nerve fibers have been visualized in the median eminence in intimate relationship with the hypophysial portal vessels. These fibers loose their fluorescence in reserpine-treated animals (FUXE, 1964). The cell bodies of the neurons may be located in the arcuate nuclei or in the ventral portions of the anterior periventricular nuclei of the hypothalamus, the region where the tubero-infundibular tract takes its origin (DAHLSTRÖM and FUXE, 1966). An LH-releasing factor could cause a discharge of these cells resulting in a release of catecholamines in the median eminence which in turn would act on the pituitary glandular cells.

Although these studies on the rat are of great interest, it remains to be seen if the results apply to other species. Reserpine apparently also inhibits ovulation in the guinea pig, but whether this is mediated by the same action of the hypothalamic-hypophysial pathways has yet to be determined (DEANESLY, 1966).

IV. Pharmacology: Effects of epinephrine and norepinephrine

The actions of epinephrine and norepinephrine, like those of adrenergic nerve stimulation, on the female reproductive organs depend upon the species of animal, hormonal state and the presence or absence of pregnancy. The published reports on the effects of the two sympathomimetic amines on the female reproductive organs are often contradictory and confusing to read let alone to interpret. One of the reasons for this is the failure of many investigators to control and standardize the experimental methodology. Some of the experimental parameters which are not always controlled include: the hormonal status of the animal, type of anesthesia, rate and route of administration of drugs, the cardiovascular effects of the administered agents. All of these factors, especially the last, are of particular importance in studies on the effects of epinephrine and norepinephrine in the intact animal. It is almost impossible to evaluate quantitatively the effects of injected epinephrine and norepinephrine on the reproductive organs *in vivo* because of the accompanying cardiovascular actions of these agents. This difficulty has prompted many investigators to use excised organs or strips thereof suspended in physiological salt solution in an organ bath. As a plea for standardization of the methods used in such studies, some points about methods will be mentioned briefly.

A. Methods of study

Since the response of the reproductive organs usually depends upon the stage of the estrous cycle or of pregnancy, the reproductive status of the experimental animal should always be stated, and controlled whenever possible by selection of animals at specific times during the estrous cycle or pregnancy. The ionic composition, temperature and pH of the physiological salt solution bathing the excised tissue should approximate that of the animal's plasma and should not be modified according to the whims of the investigator. Spontaneous motility, especially in experiments involving the isolated uterus, is a common problem in the quantitative evaluation of the actions of drugs on this organ. In an effort to reduce motility many workers lower the temperature and/or change the composition of the bathing medium (usually with respect to its calcium and magnesium concentrations). Both of these procedures can modify the response of the tissue to the drug being investigated, making the results from different laboratories difficult to compare. A solution to the problem of spontaneous motility resides in quantifying it, not suppressing it. When the contractions of the uterine muscle are being used as the response, they can be quantified by integrating the area under the recorded contraction curve for a set, standard period of time. The response of the muscle in the presence of the drug (at a standardized exposure time) can then be expressed in terms of changes in this contraction area (BISSETT et al., 1966; SULLIVAN and MARSHALL, 1969).

Patterns of spontaneous motility can change from time to time even in the absence of drugs. Such changes usually are most prominent during the first hour after isolation of the tissue, a period during which the ionic composition of the cells is equilibrating with that of the bathing fluid (DANIEL and ROBINSON, 1960; KAO, 1967). Therefore, a period of equilibration of at least one hour should be standard for all isolated organ experiments.

Much has been written about the choice of method for recording muscle contractions in isolated organ experiments, i.e. the isotonic vs. isometric techniques (CSAPO and GOODALL, 1954; SCHOFIELD, 1955; FREUND et al., 1963). Only isometric recording, or isotonic recording with optimum load, gives a quantitative estimate of the working capacity of the muscle, expressed in terms of the amount of tension developed during a contraction. With smooth muscles, as with striated, the amount of tension produced is a function of muscle length. CSAPO and GOODALL (1954) have shown that uterine muscle segments, *in vitro*, develop maximum tension at resting length. Therefore, a quantitative evaluation of the force of contraction can only be made if the muscle remains at its resting length throughout the experiment, since the same segment will develop different tensions if its length is altered. Resting length in this connection is defined as the length at which the muscle does not develop any resting tension but at which it is extended (stretched) sufficiently so that any active tension is immediately recorded. This resting length is achieved by stretching the muscle intermittantly during the initial equilibration period until a very small (about 0.5 g) resting tension is recorded. The stretch is then reduced slightly to produce zero tension, and then this length is taken as "resting length". After equilibration, the resting length usually remains remarkably constant in most isolated segments of uterine muscle, and quantitative evaluation of the changes in active tensions due to drug effects can then be made.

Isotonic recordings with a light lever and little or no load on the muscle are useful in the construction of dose-response curves particularly for bioassays. Under these conditions no contraction will be recorded until the muscle develops enough force to lift the lever. Then, because the lever is virtually weightless, very little additional force is needed to take the lever to the top of its travel. Therefore, the contraction of only a few muscle fibers will be manifest in a rapid shortening of the muscle, and the slope of the resultant dose-response curve is quite steep. A small change in concentration of the drug will produce a marked change in response, and the effects of the drug can be observed over a very narrow range of concentrations. However, when studying the quantitative relationship between the contractile tension and drug effect it is preferable to work over as wide a range of concentrations as possible. This can be done only by measuring the isometric force rather than the isotonic shortening.

When the electrical activity of the muscle is being monitored, the choice of technique depends upon the information desired. Multiple extracellular electrodes are best for measuring velocity of conduction, for indicating abnormalities of conduction including block, and for indicating the direction of spread of excitation over the muscle. The sucrose gap method is a useful extracellular technique for recording changes in membrane and action potentials for a period of many minutes from a population of cells. A modification of this method has recently been devised so that segments of uterine smooth muscle can be subjected to voltage clamp (Anderson, 1969). This should be a powerful tool for the study of ionic conductance changes during drug action. Intracellular microelectrode recordings from individual cells are useful for quantitative studies on transmembrane potential alterations produced by various drugs and agents. If this technique is used, however, a sufficient number of individual cells must be sampled to permit a statistical evaluation of the results. The sucrose gap and microelectrode techniques have been the ones most frequently employed in the modern studies of the actions of epinephrine and norepinephrine at the cellular levels (Burnstock and Holman, 1966b).

In the following account of the effects of epinephrine and norepinephrine on the female reproductive organs, only those experiments will be considered in which at least some of the requirements concerning methodology have been satisfied. The experiments were carried out primarily on the smooth muscle of the uterus and the oviduct.

B. Uterus

In the introduction to the section on the uterus in his 1933 review, Gruber states (p. 575), "Certainly no single organ in the body has been studied as thoroughly by experimenters with more conflicting results and opinions . . .". Unfortunately, what was true in 1933 is still true in 1969, particularly with regard to the actions of the sympathomimetic amines. The reason for the continued confusion is due primarily to the lack of standardization of the experimental methods. The results summarized in Table 3 are selective rather than comprehensive. The selection was made on the basis of the reliability of the experimental methods, including the control of endocrine state of the animals, composition and temperature of bathing solutions for isolated organ experiments, and notation of accompanying cardiovascular changes for the *in vivo* studies. A more complete summary of the early studies is found in Gruber (1933) and in Reynolds (1965). Miller (1967) presents an excellent summary of the more recent work particularly that with agents blocking the adrenoceptive sites.

In general the presence of both alpha and beta receptors in the myometrium of all species examined thus far has been confirmed by (1) the use of specific adrenoceptor-blocking agents, (2) the qualitative actions of epinephrine and norepinephrine, and (3) the ranking of sympathomimetic agonists according to their potencies (MILLER, 1967). The ability of the ovarian hormones estrogen and progesterone and of pregnancy to determine the relative dominance of these receptors has now been established in a number of species by studies with epinephrine and norepinephrine in the presence of appropriate adrenoceptor-blocking agents. The results of such studies, in general, substantiate the previous results with adrenergic nerve stimulation. This is particularly true for the cat, a species where "pregnancy reversal" is very striking. The uterus from non-pregnant or estrogen-dominated cats has a predominance of beta (inhibitory) receptors while that from the pregnant or progesterone-treated animal has primarily alpha (excitatory) receptors. In 1960, GRAHAM and GURD found that an aqueous extract of a pregnant or progesterone-proliferated uterus in the cat could change the normal inhibitory response of the estrogen-dominated uterus to a contraction. The effect was prevented by phenoxybenzamine and hence was due to the activation of alpha receptors. Furthermore, if a uterine segment from a progesterone-treated or pregnant cat was suspended in the same organ bath with an estrogen-treated one, there was a complete reversal of the response of the estrogenized muscle to epinephrine and norepinephrine. This interesting finding has been confirmed by TSAI and FLEMING (1964). It implies that something is present in a pregnant or progesterone-proliferated cat uterus which can be extracted by water and which can diffuse in an isolated organ bath. The substance is not found in other organs and does not behave like progesterone or oxytocin. GRAHAM and GURD (1960) suggested it may be a water-soluble, quick-acting form of pregnancy hormone or perhaps even a labile form of the adrenoceptor. The latter suggestion is provocative, and, if correct, the progesterone-dominated uterus of the cat may represent a tissue from which adrenoceptive substance could be isolated. Unfortunately, no one has attempted to analyze the aqueous extract of this tissue with modern chemical techniques.

The reversal in response (excitation to inhibition) of the intact rabbit uterus to intravenous injections of epinephrine or norepinephrine by a progestational agent has been demonstrated by WILLEMS and DE SCHAEPDRYVER (1966). These results in the intact rabbit substantiate the previous findings for the action of hypogastric nerve stimulation on the isolated uterus of the rabbit (MILLER and MARSHALL, 1965).

The influence of the ovarian hormones on the uterine response to exogenous norepinephrine in the rat is similar to that in the rabbit, i.e. an estrogen-dominated muscle is excited (alpha adrenoceptive) while a progesterone-treated one is inhibited (beta adrenoceptive) (DIAMOND and BRODY, 1966c; TOTHILL, 1967; PATON, 1968). Similar although less clear-cut effects have been reported for the guinea pig (BALASSA, 1941; HERMANSEN, 1961).

Furthermore, when the uterus from an immature guinea pig remained in an isolated organ bath for about eight hours at 37°C or was stored overnight

Table 3. *Effect of epinephrine (E) and norepinephrine (NE) on uterine motility*[a]

Animal	Condition	E	NE	Adrenoceptor blocker name	effect		Reference
Rabbit in vitro	mature, nonpregnant	↑↑		Ergot extract	abolishes	↑	DALE (1906), GADDUM (1926)
in vitro	immature	↑↑	↑	DHE	abolishes	↑	GREEFF and HOLTZ (1951)
in vitro	immature and estrogen	↑↑	↑	DHE	abolishes	↑	GREEFF and HOLTZ (1951)
in vitro	term pregnant	↑↑	↑	DHE	abolishes	↑	GREEFF and HOLTZ (1951)
in vivo	immature and estrogen	↑↑	↑	Phenoxybenzamine	reverses	↑ to ↓	WILLEMS and DE SCHAEP-DRYVER (1966)
in vivo	immature and estrogen and progestational agent[b]	↓↓	slight ↓	Propranolol	reverses	↓ to ↑	WILLEMS and DE SCHAEP-DRYVER (1966)
in vitro	mature, nonpregnant	↑↑		Dibenamine	reverses	↑ to ↓	NICKERSON and GOODMAN (1947)
in vitro	mature, nonpregnant	↑↑		Phentolamine	reverses	↑ to ↓	GROSS et al. (1951)
Guinea pig in vitro	immature	↓↓	↓				BALASSA (1941), GREEFF and HOLTZ (1951), MILLER (1967), DAVIDSON and IKOKU (1966)
in vitro	immature fresh	↓↓	↑↑	DCI	reverses	↓ to ↑	HERMANSEN (1961)
	6—8 hrs at 37° C	↑↑	↑↑				
	18 hrs at 4° C	↑↑	↑↑				

in vitro	immature 2 hrs at 37° C	↑			↑	BALASSA (1940)
in vitro	immature and gonado-trophin	↑↓	DHE	abolishes	↑	GREEFF and HOLTZ (1951)
in vitro	virgin and estradiol (100 μg)	↑↓				BALASSA (1940)
in vitro	immature and estradiol (1 μg/g)	↑↑				BALASSA (1940)
in vitro	virgin and estradiol and progesterone	→			→	HERMANSEN (1961)
in vitro	pregnant parturient	↑↓	DHE	abolishes	↑	GREEFF and HOLTZ (1951)
in vitro	virgin or pregnant	↑↓				BÜLBRING et al. (1968)
in vitro	mature diestrus	→	DHE DCI	abolishes abolishes	↑ →	CLEGG (1962)

Rat

in vitro	mature	→	Pronethalol DCI Phentolamine	abolishes abolishes no effect	→ →	LEVY and TOZZI (1963)
in vitro	mature and estrogen	→	DCI	reverses	↓ to ↑	JENSEN and VENNEROD (1961)
in vitro	mature and estrogen	→	Phentolamine DHE	prevents prevents	→ →	RUDZIK and MILLER (1962b)
in vitro	mature	→	MJ-1999	reverses	↓ to ↑	BROOKS et al. (1965)
in vitro	mature and spayed	→	Propranolol	abolishes	→	DIAMOND and BRODY (1966c)

4*

Table 3 (Continued)

Animal	Condition	E	NE	Adrenoceptor blocker name	effect		Reference
in vitro	spayed and estrogen	↓	↑	Propranolol	potentiates	↑	DIAMOND and BRODY (1966c)
				Phentolamine	reverses	↓to↑ / →to↓	DIAMOND and BRODY (1966c)
					potentiates		
					reverses	↑to↑	
in vitro	spayed and estrogen and progesterone	↓	→	Propranolol	reverses	↓to↑	DIAMOND and BRODY (1966c)
in vitro	mature and estrogen	↓	↑	Phentolamine	abolishes	↑	TOTHILL (1967)
in vitro	mature and estrogen	↓	↑	Propranolol	reverses	↓to↑	PATON (1968)
					potentiates	↑	
in vitro	pregnant, parturient	↓	↑				GREEFF and HOLTZ (1951)
in vitro	pregnant, near term	↓	↑	Phentolamine	reverses	↑to↓	MARSHALL (1967)
				Propranolol	reverses	↓to↑	
Cat							
in vivo	mature, nonpregnant	↓		Ergot extract	abolishes	→	DALE (1906), CUSHNY (1906), BÜLBRING et al. (1968)
	pregnant, early	↓					
	pregnant, late	↑↓					
in vitro	mature, virgin	↓↓	↓	DCI	reverses	↓to↑	TSAI and FLEMING (1964), VOGT (1965)
				Phentolamine	no effect		
				Phenoxybenzamine	no effect		
in vitro	pregnant, early	↑	↑	Phenoxybenzamine	reverses	↑to↓	TSAI and FLEMING (1964)
in vitro	mature, estrogen	↓↓	→↓	DCI	reverses	↓to↑	TSAI and FLEMING (1964), GRAHAM and GURD (1961)
	progesterone	↑	↑	DCI	no effect		
				Phenoxybenzamine	abolishes	↑	

	extract of pregnant or progesterone-treated uterus	DCI Phenoxybenzamine	no effect abolishes		References
Human in vivo	pregnant, mid or early	↑ or ↑↓	DCI Phenoxybenzamine	no effect abolishes	↑ KAISER and HARRIS (1950), STROUP (1962), POSE et al. (1962), BARDEN and STANDER (1968)
in vivo	in labor	↓			ESKES et al. (1965)
in vivo	pregnant, at term	↓			↑ GARRETT (1954), CIBILS et al. (1962), BARDEN and STANDER (1968)
in vivo	nonpregnant / proliferative / secretory	↑ / ↑ / ↓	Propranolol	reverses	↓ to ↑ / ↑ GARRETT (1955a), WANSBROUGH et al. (1967)
in vitro	pregnant / mid / term	↑ / ↑ / ↑			↑ / ↑ GARRETT (1955b), ADAIR and HAUGEN (1939), LEHRER (1965)
in vitro	nonpregnant / proliferative / secretory	↑ / ↑ / ↑	Phentolamine Phentolamine	abolishes abolishes	↑ ↑ MILLER et al. (1937), WANSBROUGH et al. (1967)

[a] Arrow pointing upward: increase, downward: decrease.
[b] Allyl-estrenol.
DHE = dihydroergotamine methanesulfate; DCI = 1-(3',4'-Dichlorophenyl)-2-isopropylaminoethanol hydrochloride; MJ-1999 = dl 4-(2-isopropylamino-1-hydroxyethyl) methanesulfonanilide.

at 4°C its response to epinephrine which was initially inhibitory reverses to excitation (Balassa, 1940; Hermansen, 1961). These changes suggest that the relative dominance of the alpha receptors over the beta receptors has been increased. Similar experiments apparently have never been tried for other species. However, these findings emphasize the importance of stating and standardizing the equilibration times and temperatures in studies with isolated organs.

C. Oviduct

An excellent review of the various factors influencing the motility of the oviduct has recently been published by Pauerstein et al. (1968). These authors report that, in all species ranging from the domestic hen to the human and including the pig, rabbit and monkey, the smooth muscle of the oviduct is stimulated by epinephrine. In the rabbit, the response to epinephrine varies both with the ovarian cycle and with the portion of the tube studied. Kok (1927) found that, if maturing follicles were present in the ovary, epinephrine caused a relaxation of the ampulla and a contraction of the isthmus. After ovulation this contraction of the isthmus was even more pronounced and the ampulla also gave a weak contraction. These findings were confirmed later by two Japanese workers, Murakami and Fujita (1937), but have never been extended to include the effects of adrenergic blocking agents. Brundin (1964) also noted that norepinephrine increased both the frequency and amplitude of the spontaneous contractions of the isolated oviduct of the rabbit, although he did not mention any relationship of this effect to the ovarian cycle. It will be recalled that Brundin (1965) also found that the norepinephrine content and the adrenergic innervation was greater in the isthmus than in the ampullary region of the rabbit oviduct. Both epinephrine and norepinephrine apparently increase the rate and force of contractions of the human oviduct regardless of the phase of the sexual cycle or of the portion of the tube (Sandberg et al., 1960; Hawkins, 1964).

The observations of the differential sensitivity to epinephrine in the isthmus and ampulla in the rabbit oviduct and the variations in the response to this amine with the ovarian cycle are of considerable interest. These findings in the rabbit, taken together with the regional differences in the adrenergic innervation of the oviduct in this animal, may have important functional significance in the transport of ova. Of course, other species must be studied before any general statement can be made about the physiological implications of the phenomena seen in the rabbit. A comparative study of the effects of epinephrine and norepinephrine and of the adrenergic innervation of the oviduct in a variety of species might help to elucidate the mechanism of the arrest of the ovum at the ampullary-isthmic junction that has been noted in all mammalian species examined thus far (Pauerstein et al., 1968).

D. Mode of action of epinephrine and norepinephrine at the cellular level

Most of the investigations on the effects of epinephrine and norepinephrine at the cellular level have been done on isolated segments of myometrium while the smooth muscle of the oviduct, vagina and cervix have been relatively neglected. Myometrial muscle has been the preparation of choice for a variety of reasons. First, segments of myometrium have been relatively easy to isolate from the longitudinal muscle layers in the uterine cornu of rats, rabbits and guinea pigs, especially when the animals are at the end of pregnancy or are pretreated with estrogen. Such segments exhibit regular, spontaneous contractions for many hours in the isolated organ bath. The relatively large size and uniform arrangement of the myometrial cells facilitate measurement of electrical activity with intracellular microelectrodes or with the sucrose gap. Since the bundles of muscle fibers are arrayed in a more or less longitudinal manner, in the same plane, they can be uniformly stretched until a stabile resting tension is attained. Segments of myometrium containing such a regular arrangement of muscle fibers are not easy to isolate from other species including the human and the monkey. The primate myometrium has also been unusually difficult to study with microelectrodes or with the sucrose gap methods (KUMAR et al., 1965) probably because of the complex arrangement of the muscle cells and connective tissue elements. Thus most of the subsequent discussion concerns the myometrium of the rat, rabbit and guinea pig.

Epinephrine and norepinephrine could modify the activity of the myometrial cell by an action at one or all of at least four sites; i.e. the excitable membrane, the excitation-contraction coupling mechanism, the contractile proteins, or cellular metabolism. Drugs which initiate, inhibit or stimulate the spontaneous motility of smooth muscles are most probably acting initially on the excitable membrane of the muscle cell (BURNSTOCK and HOLMAN, 1966b; GILLESPIE, 1966). The excitatory effects of epinephrine and norepinephrine on the membrane potentials of the myometrial cell are similar to those of other smooth muscle stimulants and consist of a transient depolarization of the cell membrane which either initiates an action potential discharge in the quiescent muscle or increases the frequency of this discharge in a spontaneously contracting muscle. As a result of this increase in spike frequency, the force of contraction is augmented. High doses of epinephrine and norepinephrine may produce an initial burst of spikes followed by a sustained depolarization of the cell membrane. Contraction is sustained throughout the period of depolarization (BURNSTOCK and HOLMAN, 1966b; MARSHALL, 1967).

The ionic mechanisms underlying these excitatory effects are not known. It has been suggested (BURNSTOCK and HOLMAN, 1966b) that all stimulatory drugs cause an increase in the resting conductances (permeabilities) for sodium, potassium and possibly chloride ions across the cell membrane. The overall result of such a generalized change in ionic conductance would cause a depolarization of the cell membrane. This suggestion is made by analogy with what is known to happen during the depolarization of the skeletal muscle endplate. The ionic conductances have never been measured in smooth muscle either with the muscle at rest or during the action of stimulatory drugs because of the technical difficulties encountered when such measurements are attempted in these small cells of uncertain dimensions.

The inhibitory effects of norepinephrine and epinephrine on the myometrium are associated with a reduction in frequency of spike discharge. Little or no change in the resting membrane potential occurs until doses of the amines are reached which abolish spontaneous motility. Then, there is a marked hyperpolarization of the membrane (MARSHALL, 1959; CSAPO and KURIYAMA, 1963; KUMAR et al., 1965; DIAMOND and MARSHALL, 1969a and b). A number of attempts have been made to deduce indirectly the ionic mechanisms responsible for these inhibitory effects by observing the changes in response to these amines in muscles exposed to different ionic environments (CSAPO and KURIYAMA, 1963; BURNSTOCK and HOLMAN, 1966b). The hyperpolarization is apparently related to the concentration of potassium ions in the extracellular fluid since it is increased in a K-free medium and reduced in K concentrations greater than 30 mM. Changes in extracellular sodium or chloride concentrations are usually without an effect. These findings led to the suggestion that the inhibitory actions of epinephrine and norepinephrine were related to a specific increase in potassium conductance (MARSHALL, 1968; BÜLBRING et al., 1968). However, the possibility of the activation of an electrogenic sodium pump, similar to that postulated for intestinal muscle, should not be discounted (GILLESPIE, 1966).

A technique has recently been described (ANDERSON, 1969) whereby small segments of myometrium can be subjected to voltage clamp, so that it should now be feasible to determine directly the specific membrane conductances for sodium, potassium, chloride and other ions under the normal conditions and during drug action. It will then be possible for the first time to work out the details of drug action at the membrane level in terms of changes in these conductances. The results of such experiments will undoubtedly represent major milestones in our understanding of the effects of drugs, hormones and neurotransmitters on the myometrium at the cellular level.

The changes in membrane potentials accompanying the stimulatory and inhibitory effects of epinephrine and norepinephrine in the rat uterus are reversed in the presence of alpha receptor- and beta receptor-blocking agents respectively (MARSHALL, 1967; DIAMOND and MARSHALL, 1969a). If we assume that alterations in the ionic permeability of the cell membrane underlie the membrane potential changes, then effects upon the alpha and upon the beta receptor are both mediated via these permeability changes.

BÜLBRING et al. (1968) propose that the cellular mechanism underlying the phenomenon of pregnancy reversal in the cat uterus might be related to specific changes in the ionic permeability caused by epinephrine during the gravid vs. the non-gravid state. In the virgin cat the chloride content of the myometrium is apparently quite low, but in early pregnancy it is nearly doubled. Thus, in the pregnant uterus, the chloride equilibrium potential is considerably below the membrane potential of the myometrial cell. Therefore, if epinephrine caused a selective increase in the membrane permeability to chloride, it would depolarize the cell and have an excitatory effect. On the other hand, in the virgin uterus, if epinephrine produced a selective increase in potassium permeability (which apparently it does), the membrane would hyperpolarize and inhibition would result. Although these changes in permeability have not been measured directly, the experiments of BÜLBRING and her colleagues provide the first insight into the possible cellular mechanisms for the reversal in the response to epinephrine and norepinephrine that is so characteristic of the uterus of various species. It is conceivable that the adrenoceptive sites regulate the ionic permeability of the myometrial cell and in this manner control the direction of the response to epinephrine and norepinephrine.

The effects of epinephrine and norepinephrine are not limited to the excitable membrane, although under normal conditions one would assume that their initial effects are mediated through alterations in the electrical activity of the membrane. Nevertheless, uterine muscles, whose membranes have been depolarized by isotonic potassium solutions to a level where all action potentials are abolished, still will relax (rat) or contract (rabbit) in response to epinephrine and norepinephrine. These actions are prevented by the appropriate adrenoceptor-blocking agents indicating that they are mediated by the same receptors as in the normally polarized muscles (EDMAN and SCHILD, 1963; SCHILD, 1967). Since these effects on the depolarized muscles are intimately related to the calcium concentration in the bathing medium, it has been suggested that the site of action may be on excitation-contraction coupling (EDMAN and SCHILD, 1963). No results have been reported for the action of epinephrine and norepinephrine on the contractile elements of the myometrium.

It is well established that epinephrine stimulates metabolic processes generating energy-rich phosphate compounds in a variety of tissues including the smooth muscle of the intestine (BUEDING et al., 1967). This is apparently an effect upon beta receptors and at one time it was implicated as the energy source for the membrane hyperpolarization accompanying the action of epinephrine in the intestine (BUEDING and BÜLBRING, 1964). This idea has now been modified to include the excitatory effects of epinephrine as well (BÜLBRING et al., 1968). In the rat uterus, epinephrine also stimulates energy metabolism as indicated by its ability to increase phosphorylase activity. This metabolic action is also blocked by beta receptor blockers (Diamond and Brody, 1966a and b). However, the relationship between the metabolic action and the contractile effects or membrane effects is unclear.

V. Concluding remarks

When this review was begun, it was hoped that at its completion some general statements could be made about the overall role of the peripheral adrenergic nerves in the control and regulation of the female reproductive organs. As the preparation of the review progressed, it became increasingly apparent that such generalizations could not be made. Although it has been known for some time that stimulation of the hypogastric nerves causes the uterus to contract or to relax depending upon the appropriate species and the endocrine state of the individual, the precise role of the adrenergic nerves, when stimulated at physiological frequencies, in the functional activity of the uterus has not been identified. The importance of the efferent innervation of the uterus, vagina and oviduct is not as immediately obvious as is that of skeletal or cardiac muscle or of some smooth muscles, e.g. the iris, the nictitating membrane. In the reproductive organs the endocrine regulation has always been more apparent and therefore more emphasized than the neural control, especially at peripheral sites. However, the complex neuro-endocrine interactions at peripheral effector sites are just beginning to be recognized. We have seen examples of these interactions: the changes in the effective innervation density and in transmitter content of the adrenergic nerves and the dependence of such changes on pregnancy and on estrogen administration; the modification by ovarian hormones and during pregnancy of the adrenoceptors on the myometrial cell; the change in uterine sensitivity to oxytocin during hypogastric nerve stimulation. Nevertheless, the interrelations between the endocrines and the nervous system in the integration of the total activity pattern of the female reproductive tract have yet to be elucidated. It is conceivable that neuro-endocrine coordination of muscular activity might be needed for (1) transport of sperm in the female tract, (2) transport and spacing of embryos before implantation, (3) maintenance of pregnancy, and (4) parturition.

The female reproductive organs, of which the uterus has been the most extensively studied, are under a hierarchical system of control. For example, at the lowest level is the myogenic activity of the myometrium, the fundamental nature of which is exemplified by intrinsic pacemaker activity and the myogenic conduction of excitation. Myogenic activity can be modified locally by factors such as stretch or by circulating hormones. At the next level operate the spinal reflex mechanisms such as those initiated by distention of the vagina. These, in turn, may modulate the intrinsic, myogenic activity either directly by exciting or inhibiting the muscle cells, or indirectly by changing their sensitivity to circulating hormones. At the hypothalamic level there are additional controlling mechanisms discharging through both the adrenergic and hypophysial pathways. Above these are undoubtedly higher centers in the cerebral cortex that may exercise control by relays through the hypothalamus and spinal pathways.

The present review has focused primarily on the peripheral sites. It is apparent that we are just beginning to learn something about the myogenic activity of the myometrium and how it may be influenced by neurohumoral agents. But most of our information is descriptive and concerns processes several steps removed from the fundamental ones. For example, we still speak in vague terms about "adrenoceptors" and about the way in which epinephrine and norepinephrine exert their actions at the cellular level. Our knowledge of the morphology and ultrastructure of the intrinsic innervation of the reproductive tract is just beginning to unfold and has not yet been correlated with the physiological effects of nerve stimulation at the level of the organ, the tissue, or the cell.

One may envy the future reviewer who can summarize the role of the adrenergic nervous system in the reproductive processes of the female by describing with some exactness in a variety of species: (1) the identity and activity of nerve cells and pathways controlling the release of the neuro-transmitters, (2) the factors influencing neuro-transmission, (3) the specific changes of the effector cells caused by the neuro-transmitter and by circulating hormones, and (4) the nature of the adrenoceptor. The journey to this goal may be arduous and long. Hopefully, it will be aided and hastened by the advent and application of new approaches and techniques.

References

ABRAHAMS, V. C., LANGWORTHY, E. P., THEOBALD, G. W.: Potentials evoked in the hypothalamus and cerebral cortex by electrical stimulation of the uterus. Nature (Lond.) 203, 654—656 (1964).

ADAIR, F. L., HAUGEN, J. A.: A study of suspended human uterine muscle strips *in vitro*. Amer. J. Obstet. Gynec. 37, 753—761 (1939).

ADRIAN, E. D., BRONK, D. W., PHILLIPS, G.: Discharges in mammalian sympathetic nerves. J. Physiol. (Lond.) 74, 115—133 (1932).

AHLQUIST, R. P.: A study of the adrenotropic receptors. Amer. J. Physiol. 153, 586—600 (1948).

— The adrenergic receptor. J. pharm. Sci. 55, 359—370 (1966).

ANDERSON, L. L., BOWERMAN, A. M., MELAMPY, R. M.: Neuro-utero-ovarian relationships. In: Advances in Neuroendocrinology. Urbana: Univ. Illinois Press 1963.

ANDERSON, N. C.: Voltage-clamp studies of uterine smooth muscle. J. gen. Physiol. 54, 145—166 (1969).

BALASSA, G.: Uterine potentials. J. Pharmacol. exp. Ther. 70, 189—200 (1940).

BARDEN, T. P., STANDER, R. W.: Effects of adrenergic blocking agents and catecholamines in human pregnancy. Amer. J. Obstet. Gynec. **102**, 226—235 (1968).

BARNEA, A., SHELESNYAK, M. C.: The catecholamine content of the uterus, ovary and hypophysis during early pregnancy. J. Endocr. **31**, 271—278 (1965).

BARRACLOUGH, C. A., CROSS, B. A.: Unit activity in the hypothalamus of the cyclic female rat: effect of genital stimuli and progesterone. J. Endocr. **26**, 339—359 (1963).

BELL, C.: Dual vasoconstrictor and vasodilator innervation of the uterine arterial supply in the guinea pig. Circulat. Res. **23**, 279—289 (1968).

BERGMAN, R. A.: Uterine smooth muscle fibers in castrate and estrogen-treated rats. J. Cell Biol. **36**, 639—648 (1968).

BERTLER, A., CARLSSON, A., ROSENGREN, E.: A method for the fluorimetric determination of adrenaline and noradrenaline in tissues. Acta physiol. scand. **44**, 273—292 (1958).

BISSETT, G. W., HALDAR, J., LEWIN, J. E.: Actions of oxytocin and other biologically active peptides on the rat uterus. Mem. Soc. Endocr. **14**, 185—204 (1966).

BOWER, E. A.: The characteristics of spontaneous and evoked action potentials recorded from the rabbit's uterine nerves. J. Physiol. (Lond.) **183**, 730—747 (1966a).

BOWER, E. A.: The activity of post-ganglionic sympathetic nerves to the uterus of the rabbit. J. Physiol. (Lond.) **183**, 748—767 (1966b).

BOWMAN, W. C., NOTT, M. W.: Actions of sympathomimetic amines and their antagonists on skeletal muscle. Pharmacol. Rev. **21**, 27—72 (1969).

BRONK, D. W., TOWER, S. S., SOLDANT, D. Y., LARRABEE, M. G.: The transmission of trains of impulses through a sympathetic ganglion and in its postganglionic nerves. Amer. J. Physiol. **122**, 1—15 (1938).

BROOKS, J. R., SCHAEPPI, U., PINCUS, G.: Evidence for the presence of alpha adrenergic excitatory receptors in the rat uterus. Life Sci. **4**, 1817—1821 (1965).

BRUNDIN, J.: An occlusive mechanism in the fallopian tube of the rabbit. Acta physiol. scand. **61**, 219—227 (1964).

— Distribution and function of adrenergic nerves in the rabbit fallopian tube. Acta physiol. scand. **66**, 5—57 (1965).

BUEDING, E., BÜLBRING, E.: The inhibitory action of adrenaline. In: Pharmacology of Smooth Muscle. Proc. 2nd. Int. Pharmacol. Meeting, vol. 6, p. 37—56. Oxford: Pergamon 1964.

— — GERKEN, G., HAWKINS, J. T., KURIYAMA, H.: The effect of adrenaline on the adenosine tri-phosphate and createnine phosphate content of intestinal smooth muscle. J. Physiol. (Lond.) **193**, 197—212 (1967).

BÜLBRING, E., CASTEELS, R., KURIYAMA, H.: Membrane potential and ion content in cat and guinea-pig myometrium and the response to adrenaline and noradrenaline. Brit. J. Pharmacol. **34**, 388—407 (1968).

BURDICK, H. O., PINCUS, G.: The effect of estrin injections upon the development of ova in mice and rabbits. Amer. J. Physiol. **111**, 201—208 (1935).

BURNSTOCK, G., CAMPBELL, G., BENNETT, M., HOLMAN, M. E.: Inhibition of the smooth muscle of the taenia coli. Nature (Lond.) **200**, 581—582 (1963).

— HOLMAN, M. E.: The transmission of excitation from autonomic nerve to smooth muscle. J. Physiol. (Lond.) **155**, 115—133 (1961).

— — Junction potentials at adrenergic synapses. Pharmacol. Rev. **18**, 481—493 (1966a).

— — Effect of drugs on smooth muscle. Ann. Rev. Pharmacol. **6**, 129—156 (1966b).

— — PROSSER, C. L.: Electrophysiology of smooth muscle. Physiol. Rev. **43**, 482—527 (1963).

— ROBINSON, P. M.: Localization of catecholamines and acetylcholinesterase in autonomic nerves. Circulat. Res., Suppl. III, **21**, 43—55 (1967).

CAESAR, R., EDWARDS, G. A., RUSKA, H.: Architecture and nerve supply of mammalian smooth muscle tissue. J. biophys. biochem. Cytol. **3**, 867—878 (1957).

CALDEYRO-BARCIA, R., ALVAREZ, H.: Effect of presacral nerve stimulation on the contractility of the nonpregnant human uterus. J. appl. Physiol. **6**, 556—558 (1954).

CANNON, W. B., BACQ, Z. M.: Studies on the conditions of activity in endocrine organs. XXVI. A hormone produced by sympathetic action on muscle. Amer. J. Physiol. 96, 392—412 (1931).

CARLSSON, A., FALCK, B., HILLARP, N. A.: Cellular localization of monoamines. Acta physiol. scand., Suppl. 196, 56, 1—28 (1962).

— ROSENGREN, E., BERTLER, Å., NILSSON, J.: Effect of reserpine on the metabolism of catecholamines. In: Psychotropic Drugs. Amsterdam: Elsevier Publ. Co. 1957.

CELANDER, O.: The range of control exercised by the sympathico-adrenal system. Acta. physiol. scand. 32, Suppl. 116, 1—132 (1954).

CHA, K. S., LEE, W. C., RUDZIK, A., MILLER, J. W.: A comparison of the catecholamine concentrations of uteri from several species and the alterations which occur during pregnancy. J. Pharmacol. exp. Ther. 148, 9—13 (1965).

CIBILS, L. A., POSE, S. V., ZUSPAN, F. P.: The effect of 1-norepinephrine infusion on uterine contractility and cardiovascular system. Amer. J. Obstet. Gynec. 84, 307—317 (1962).

CLEGG, P. C.: The effect of adrenergic blocking agents on the guinea-pig uterus *in vitro* and a study of the histology of the intrinsic myometrial nerves. J. Physiol. (Lond.) 169, 73—90 (1962).

COPPOLA, J. A., LEONARDI, R. G., LIPMANN, W.: Ovulatory failure in rats after treatment with brain norepinephrine depletor. Endocrinology 78, 225—228 (1966).

CROSS, B. A.: The motility and reactivity of the oestrogenized rabbit uterus *in vivo;* with comparative observations on milk ejection. J. Endocr. 16, 237—260 (1958).

CROUT, J. R., CREVELING, C. R., UDENFRIEND, S.: Norepinephrine metabolism in rat brain and heart. J. Pharmacol. exp. Ther. 132, 269—273 (1961).

CSAPO, A., GOODALL, M.: Excitability, length tension relation and kinetics of uterine muscle contraction in relation to hormonal status. J. Physiol. (Lond.) 126, 384—395 (1954).

— KURIYAMA, H. A.: Effects of ions and drugs on cell membrane activity and tension in the postpartum rat myometrium. J. Physiol. (Lond.) 165, 575—592 (1963).

CUSHNY, A. R.: On the movements of the uterus. J. Physiol. (Lond.) 35, 1—19 (1906).

DAHLSTRÖM, A., FUXE, K.: Monoamines and the pituitary gland. Acta endocr. (Kbh.) 51, 301—314 (1966).

DALE, H. H.: On some physiological actions of ergot. J. Physiol. (Lond.) 34, 163—206 (1906).

— Nomenclature of fibres in the autonomic system and their effects. J. Physiol. (Lond.) 80, 10—11 (1933).

— Nomenclature of fibres in the autonomic system and their effects. In: Adventures in physiology, p. 528—529. London: Pergamon Press 1953.

— The beginnings and the prospects of neurohumoral transmission. Pharmacol. Rev. 6, 7—13 (1954).

— LAIDLAW, P. P.: The significance of the suprarenal capsules in the action of certain alkaloids. J. Physiol. (Lond.) 45, 1—23 (1912).

DANIEL, E. E., ROBINSON, K.: The secretion of sodium and uptake of potassium by isolated uterine segments made sodium-rich. J. Physiol. (Lond.) 154, 421—445 (1960).

DAVID, A., CZERNOBILSKY, B.: A comparative histologic study of the uterotubal junction in the rabbit, rhesus monkey, and the human female. Amer. J. Obstet. Gynec. 101, 417—421 (1968).

DAVIDSON, W. J., IKOKU, C.: The adrenergic receptors in the guinea-pig uterus. Canad. J. Physiol. Pharmac. 44, 491—493 (1966).

DAVIES, P. W.: The action potential. In: Medical physiology. St. Louis: C. V. Mosby Co. 1968.

DAVIS, A. A.: The innervation of the uterus. J. Obstet. Gynaec. 40, 481—497 (1933).

DEANESLY, R.: The effects of reserpine on ovulation and on the corpus luteum of the guinea-pig. J. Reprod. Fertil. **11**, 429—438 (1966).

DIAMOND, J., BRODY, T. M.: Effect of catecholamines on smooth muscle motility and phosphorylase activity. J. Pharmacol. exp. Ther. **152**, 202—211 (1966a).

— — Relationship between smooth muscle contraction and phosphorylase activation. J. Pharmacol. exp. Ther. **152**, 212—220 (1966b).

— — Hormonal alteration of the response of the rat uterus to catecholamines. Life Sci. **5**, 2187—2193 (1966c).

— MARSHALL, J. M.: Smooth muscle relaxants; dissociation between resting membrane potential and resting tension in the myometrium. J. Pharmacol. exp. Ther. **168**, 13—20 (1969a).

— — A comparison of the effects of various smooth muscle relaxants on the electrical and mechanical activity of the rat uterus. J. Pharmacol. exp. Ther. **168**, 21—30 (1969b).

EDMAN, K. A. P., SCHILD, H. O.: Calcium and the stimulant and inhibitory effects of adrenaline in depolarized smooth muscle. J. Physiol. (Lond.) **169**, 404—411 (1963).

EL-BADAWI, A., SCHENK, E. A.: The peripheral adrenergic innervation apparatus. I. Intraganglionic and extraganglionic adrenergic ganglion cells. Z. Zellforsch., Abt. Histochem. **87**, 218—225 (1968).

ERÄNKO, O.: Histochemistry of nervous tissues: Catecholamines and cholinesterases. Ann. Rev. Pharmacol. **7**, 203—222 (1967).

ESKES, T., STOLTE, L., SEELEN, J., MOLD, H. D., VOGELSANG, C.: Epinephrine derivates and the activity of the human uterus. II. The influence of pronethalol and propranolol on the uterine and systemic activity of p-hydroxphenylisopropylarterenol (Cc-25). Amer. J. Obstet. Gynec. **92**, 871—881 (1965).

EULER, U. S. VON, LISHAJKO, F.: A specific kind of noradrenaline granules in the vesicular gland and the vas deferens of the bull. Life Sci. **5**, 687—691 (1966).

EVERETT, J. W.: Central neural control of reproductive functions of the adenohypophysis. Physiol. Rev. **44**, 374—431 (1964).

FALCK, B.: Observations on the possibilities of the cellular localization of monoamines by a fluorescence method. Acta physiol. scand. **56**, 197, 1—25 (1962).

— HILLARP, N. A., THIENE, G., TORP, A.: Fluorescence of catecholamines and related compounds condensed with formaldehyde. J. Histochem. Cytochem. **10**, 348—354 (1962).

FELDBERG, W.: HENRY HALLETT DALE, 1875—1968. Brit. J. Pharmacol. **35**, 1—9 (1969).

FERGUSON, J. K. W.: A study of the motility of the intact uterus at term. Surg. Gynec. Obstet. **73**, 359—366 (1941).

FERRY, C. B.: The innervation of the vas deferens of the guinea-pig. J. Physiol. (Lond.) **192**, 463—478 (1967).

FOLKOW, B.: Impulse frequency in sympathetic vasomotor fibres correlated to the release and elimination of the transmitter. Acta physiol. scand. **25**, 49—76 (1952).

FRANKENHÄUSER, F.: Die Nerven der Gebärmutter und ihre Endigungen in den glatten Muskelfasern. Ein Beitrag zur Anatomie und Gynäkologie. Jena: Mauke 1867.

FREUND, M., WIEDERMAN, J., SAPHIER, A.: A method for the simultaneous recording, *in vitro*, of the motility of the vagina, of the body of the uterus, and of both uterine horns in the guinea pig. Fertil. and Steril. **14**, 416—430 (1963).

FUXE, E.: Cellular localization of monoamines in the median eminence and the infundibular stem of some mammals. Z. Zellforsch., Abt. Histochem. **61**, 710—724 (1964).

GADDUM, J. H.: The action of adrenalin and ergotamine on the uterus of the rabbit. J. Physiol. (Lond.) **61**, 141—150 (1926).

GAFFNEY, E., BURKET, R. L., WORONKOW, S.: Catecholamine content of the pregnant and nonpregnant human uterus. Obstet. and Gynec. **25**, 340—342 (1965).

GANSLER, H.: Phasenkontrast- und elektronenmikroskopische Untersuchungen zur Innervation der glatten Muskulatur. Acta neuroveg. (Wien) **22**, 192—211 (1960).

GARRETT, W. J.: The effects of adrenaline and noradrenaline on the intact human uterus in late pregnancy and labour. J. Obstet. Gynaec. Brit. Emp. 61, 586—592 (1954).

— The effects of adrenaline and noradrenaline on the intact nonpregnant human uterus. J. Obstet. Gynaec. Brit. Emp. 62, 876—881 (1955a).

— The effects of adrenaline, noradrenaline and dihydroergotamine on excised human myometrium. Brit. J. Pharmacol. 10, 39—42 (1955b).

GAUNT, R., CHART, J. J., RENZI, A. A.: Interactions of drugs with endocrines. Ann. Rev. Pharmacol. 3, 109—127 (1963).

GILLESPIE, J. S.: The mode of action of catecholamines on smooth muscle. Mem. Soc. Endocr. 14, 155—167 (1966).

GRAHAM, J. P. D., GURD, M. R.: Effects of adrenaline on the isolated uterus of the cat. J. Physiol. (Lond.) 152, 243—249 (1960).

GREEFF, K., HOLTZ, P.: Über die Uteruswirkung des Adrenalins und Arterenols. Ein Beitrag zum Problem der Uterusinnervation. Arch. int. Pharmacodyn. 88, 228—252 (1951).

GREEN, R. D., MILLER, J. W.: Catecholamine concentrations. Changes in plasma of rats during estrous cycle and pregnancy. Science 151, 825—826 (1966a).

— — The effect of various substances on the efflux of labeled catecholamines from the uterus of the rat. J. Pharmacol. exp. Ther. 152, 439—444 (1966b).

GREENWALD, G. S.: In vivo recording of intraluminal pressure change in the rabbit oviduct. Fertil. and Steril. 14, 665—672 (1963).

GREISS, F. C.: The uterine vascular bed: effect of adrenergic stimulation. Obstet. and Gynec. 21, 295—301 (1963).

— GOBBLE, F. L.: Effect of sympathetic nerve stimulation on the uterine vascular bed. Amer. J. Obstet. Gynec. 97, 962—967 (1967).

— GOBBLE, JR., F. L., ANDERSON, S. G., McGUIRT, W. F.: Effect of parasympathetic nerve stimulation on the uterine vascular bed. Amer. J. Obstet. Gynec. 99, 1067—1072 (1967).

GROSS, F., TRIPOD, J., MEIER, R.: Regitin (Präparat C 7337) ein neues Imidazolin-derivat mit spezifischer sympathikolytischer Wirkung. Schweiz. med. Wschr. 81, 352—357 (1951).

GRUBER, C. M.: The autonomic innervation of the genito-urinary system. Physiol. Rev. 13, 497—609 (1933).

GUNN, J. A., FRANKLIN, K. J.: The sympathetic innervation of the vagina. Proc. roy. Soc. B 94, 197—203 (1922).

GUSTAVSON, R. G., DYKE, H. B. VAN: Further observations on the pregnancy response if the uterus of the cat. J. Pharmacol. exp. Ther. 41, 139—146 (1931).

GUTMAN, Y., WEIL-MALHERBE, H.: Subcellular distribution of norepinephrine in uteri of some species. Nature (Lond.) 214, 108—109 (1967).

HARMAN, J. W., O'HEGARTY, M. T., BYRNES, C. R.: The ultrastructure of human smooth muscle. Exp. molec. Path. 1, 204—227 (1962).

HASAMA, B.: Elektrophysiologische Untersuchungen über die Uterusbewegungen. Pflügers Arch. ges. Physiol. 229, 100—112 (1931).

HAWKINS, D. F.: Some pharmacological reactions of isolated rings of human fallopian tube. Arch. int. Pharmacodyn. 152, 474—481 (1964).

HERMANSEN, K.: The effect of adrenaline, noradrenaline and isoprenaline on the guinea-pig uterus. Brit. J. Pharmacol. 16, 116—128 (1961).

HOLMAN, M. E.: Some electrophysiological aspects of transmission from noradrenergic nerves to smooth muscle. Circulat. Res., Suppl. III, 21, 71—81 (1967).

HOPKINS, T. F., PINCUS, G.: Effects of reserpine on gonadotropin-induced ovulation in immature rats. Endocrinology 73, 775—780 (1966).

JACOBOWITZ, D., WALLACH, E. E.: Histochemical and chemical studies of the autonomic innervation of the ovary. Endocrinology 81, 1132—1139 (1967).

JACOBSON, H. N., NIEVES, O.: Intrinsic nerve fibers of the primate endometrium. Exp. Neurol. **4**, 180—187 (1961).

JAEGER, J.: Electronenoptische Untersuchungen an der glatten Muskulatur des menschlichen graviden Uterus. Gynaecologia (Basel) **154**, 193—205 (1962).

JENSEN, K. B., VENNEROD, A. M.: Reversal of the inhibitory action of adrenaline and histamine on the rat uterus. Acta pharmacol. (Kbh). **18**, 298—306 (1961).

KAHANE, Z., VESTERGAARD, P.: An improved method for measurement of free epinephrine and norepinephrine with a phosphate-metaphosphate buffer in the trihydroxyinodole procedure. J. Lab. clin. Med. **65**, 848—858 (1965).

KAISER, I. H., HARRIS, J. S.: The effect of adrenaline on the pregnant human uterus. Amer. J. Obstet. Gynec. **59**, 775—784 (1950).

KAO, C. Y.: Ionic basis of electrical activity in uterine smooth muscle. In: Cellular biology of the uterus. New York: Appleton-Century Crofts 1967.

KLINGMAN, G.: Catecholamine levels and dopa-decarboxylase activity in peripheral organs and adrenergic tissues in the rat after immunosympathectomy. J. Pharmacol. exp. Ther. **148**, 14—21 (1965).

KOK, F.: Pharmacological influence on the muscles of the tube. A contribution to the question of the transport of the ovum. Zbl. Gynäk. **51**, 2650—2662 (1927).

KRANTZ, K. E.: Innervation of the human uterus. Ann. N.Y. Acad. Sci. **75**, 770—784 (1959).

KUMAR, O., WAGATSUMA, T., BARNES, A. C.: *In vitro* hyperpolarizing effect of adrenaline on human myometrial cell. Amer. J. Obstet. Gynec. **91**, 575—577 (1965).

KUSHIYA, I.: An electron microscope study of muscular coats in the ampulla of the rabbit oviduct, with special references to the neuromuscular relationship. J. Electron Micr. **17**, 127—138 (1968).

LABATE, J. S.: Influence of cocaine on the uterine reaction induced by adrenaline and hypogastric nerve stimulation. J. Pharmacol. exp. Ther. **72**, 370—382 (1941).

LAGUENS, R., LAGRUTTA, J.: Fine structure of human uterine muscle in pregnancy. Amer. J. Obstet. Gynec. **89**, 1040—1048 (1964).

LANGLEY, J. N.: On the stimulation and paralysis of nerve cells and of nerve-endings. Part 1. J. Physiol. (Lond.) **27**, 224—236 (1901a).

— Observations on the physiological action of extracts of the supra-renal bodies. J. Physiol. (Lond.) **27**, 237—256 (1901b).

— ANDERSON, H. K.: The constituents of the hypogastric nerves. J. Physiol. (Lond.) **17**, 177—191 (1894).

— — The innervation of the pelvic and adjoining viscera. Part IV. The internal generative organs. J. Physiol. (Lond.) **19**, 122—130 (1895a).

— — The innervation of the pelvic and adjoining viscera. Part V. Position of the nerve cells on the course of the efferent nerve fibres. J. Physiol. (Lond.) **19**, 131—139 (1895b).

— — The innervation of the pelvic and adjoining viscera. Part VII. Anatomical observations. J. Physiol. (Lond.) **20**, 372—406 (1896).

LEHRER, D. N.: The effect of spasmolytic agents on the isolated human myometrium. J. Pharm. Pharmacol. **17**, 584—592 (1965).

LEMPINEN, M.: Extra-adrenal chromaffin tissue of the rat and the effect of cortical hormones on it. Acta physiol. scand. **62**, Suppl. 231 (1964).

LEVY, B., TOZZI, S.: The adrenergic receptive mechanism of the rat uterus. J. Pharmacol. exp. Ther. **142**, 178—184 (1963).

LOEWI, O.: Über humorale Übertragbarkeit der Herznervenwirkung. I. Mitt. Pflügers Arch. ges. Physiol. **189**, 239—242 (1921).

MALMFORS, T.: Studies on adrenergic nerves. Acta. physiol. scand. **64**, Suppl. 248, 7—95 (1965).

MANN, M., WEST, G. B.: The nature of uterine and intestinal sympathin. Brit. J. Pharmacol. **6**, 79—82 (1951).

MARSHALL, J. M.: Effects of estrogen and progesterone on single uterine muscle fibers in the rat. Amer. J. Physiol. **197**, 936—941 (1959).

— Comparative aspects of the pharmacology of smooth muscle. Fed. Proc. **26**, 1104—1110 (1967).

— Relation between the ionic environment and the action of drugs on the myometrium. Fed. Proc. **27**, 115—119 (1968).

MEYERSON, B. J., SAWYER, C. H.: Monoamines and ovulation in the rat. Endocrinology **83**, 170—176 (1968).

MILLER, E. G., COCKERILL, J. R., KURZROK, R.: Reactions of human uterine muscle *in vitro* to pituitrin, adrenaline and acetylcholine and their relations to the menstrual cycle. Amer. J. Obstet. Gynec. **33**, 154—156 (1937).

MILLER, J. W.: Adrenergic receptors in the myometrium. Ann. N.Y. Acad. Sci. **139**, 788—798 (1967).

MILLER, M. D., MARSHALL, J. M.: Uterine response to nerve stimulation; relation to hormonal status and catecholamines. Amer. J. Physiol. **209**, 859—865 (1965).

MITCHELL, G. A. G.: The innervation of the ovary, uterine tube, testis and epididymis. J. Anat. (Lond.) **72**, 508—517 (1938).

MORISON, R. S.: The effects of adrenaline and of nerve stimulation on the mechanical and electric responses of uterine muscle. Amer. J. Physiol. **128**, 372—381 (1940).

MURAKAMI, S., FUJITA, I.: About the action of several pharmaceuticals upon the uterus and the tubes. Folia pharmacol. jap. **23**, 31—45 (1937).

NAKANISHI, H., WANSBROUGH, H., WOOD, C.: Postganglionic sympathetic nerve innervating human fallopian tube. Amer. J. Physiol. **213**, 613—619 (1967).

— WOOD, C.: Effects of adrenergic blocking agents on human fallopian tube motility *in vitro*. J. Reprod. Fertil. **16**, 21—28 (1968).

NICKERSON, M., GOODMAN, L. S.: Pharmacological properties of a new adrenergic blocking agent: N,N-Dibenzyl-β-chlorethylamine (Dibenamine). J. Pharmacol. exp. Ther. **89**, 167—185 (1947).

NILSSON, R.: Rate of depletion of noradrenaline in some peripheral tissue induced by a reserpine injection. Experientia (Basel) **20**, 679 (1964).

NORBERG, J. G., ROSENGREN, E., SJÖBERG, N. O.: On the occurrence of 5-hydroxy-tryptamine containing cells in the vaginal and vestibular epithelium of the rabbit. Z. Zellforsch., Abt. Histochem. **63**, 302—308 (1964).

NORBERG, K. A., FREDRICSSON, B.: Cellular distribution of monoamines in the uterine and tubal walls of the rat. Acta physiol. scand. **68**, Suppl. 277, 149 (1966).

— HAMBERGER, B.: The sympathetic adrenergic neuron. Some characteristics revealed by histochemical studies on the intraneuronal distribution of the transmitter. Acta physiol. scand. **63**, Suppl. 238, 5—42 (1964).

— RITZEN, M., UNGERSTEDT, U.: Histochemical studies on a special catecholamine-containing cell type in sympathetic ganglia. Acta. physiol. scand. **67**, 260—270 (1966).

ORLOV, R. S.: The intracellular recording of smooth muscle potentials during stimulation of excitatory and inhibitory nerves. J. Physiol. USSR **47**, 500—503 (1961).

OSKARSSON, V.: Influence of ovarian hormones and denervation on the catecholamines of the rat uterus. Acta. endocr. **34**, 38—44 (1960).

OWMAN, C., ROSENGREN, E., SJÖBERG, N. O.: Origin of the adrenergic innervation to the female genital tract of the rabbit. Life Sci. **5**, 1389—1396 (1966).

— — — Adrenergic innervation of the human female reproductive organs: A histochemical and chemical investigation. J. Obstet. Gynaec. **30**, 763—773 (1967).

— SJÖBERG, N. O.: Adrenergic nerves in the female genital tract of the rabbit. With remarks on cholinesterase-containing structures. Z. Zellforsch., Abt. Histochem. **74**, 182—197 (1966).

— SJÖSTRAND, N. O.: Short adrenergic neurons and catecholamine-containing cells in vas deferens and accessory male genital glands of different mammals. Z. Zellforsch., Abt. Histochem. **66**, 300—320 (1965).

Pallie, W., Corner, G. W., Weddell, G.: Nerve terminations in the myometrium of the rabbit. Anat. Rec. 118, 789—812 (1954).

Paton, D. M.: The contractile response of the isolated rat uterus to noradrenaline and 5-hydroxytryptamine. Europ. J. Pharmacol. 3, 310—315 (1968).

Pauerstein, C. J., Woodruff, J. D., Zachary, A. S.: Factors influencing physiologic activities in the fallopian tube, the anatomy, physiology and pharmacology of tubal transport. Obstet. gynec. Surv. 23, 215—243 (1968).

Pennefather, J. N., Isaac, P. F.: The response of the isolated uterus of the guinea-pig to stimulation of the ovarian nerves during the oestrous cycle. Aust. J. exp. Biol. med. Sci. 45, 229—244 (1967).

Pose, S. V., Cibils, L. A., Zuspan, F. P.: Effect of l-epinephrine infusion on uterine contractility and cardiovascular system. Amer. J. Obstet. Gynec. 84, 297—306 (1962).

Reynolds, S. M. R.: Physiology of the uterus, 2nd ed. New York: Hafner Publ. Co. 1965.

Richardson, K. C.: The fine structure of autonomic nerve endings in the smooth muscle of the rat vas deferens. J. Anat. (Lond.) 96, 427—442 (1962).

Rosengren, E., Sjöberg, N. O.: The adrenergic nerve supply to the female reproductive tract of the cat. Amer. J. Anat. 121, 271—284 (1967).

— — Changes in the amount of adrenergic transmitter in the female genital tract of the rabbit during pregnancy. Acta. physiol. scand. 72, 412—424 (1968).

Rudolph, L., Ivy, A. C.: The physiology of the uterus in labor. Amer. J. Obstet. Gynec. 19, 317—335 (1930).

Rudzik, A. D., Miller, J. W.: The mechanism of uterine inhibitory action of relaxin-containing ovarian extracts. J. Pharmacol. exp. Ther. 138, 82—87 (1962a).

— — The effect of altering the catecholamine content of the uterus on the rate of contractions and the sensitivity of the myometrium to relaxin. J. Pharmacol. exp. Ther. 138, 88—95 (1962b).

Rüsse, M., Marshall, J. M.: Effect of hypogastric nerve stimulation on the uterus of the guinea pig in vivo. Fed. Proc. 29, 639 (1969).

Sandberg, F., Ingelman-Sunberg, A., Lindgren, L., Ryden, G.: In vitro studies of the motility of the human fallopian tube. Acta obstet. gynec. scand. 39, 506—516 (1960).

Sauer, J., Jett-Jackson, C. E., Reynolds, S. R. M.: Reactivity of the uterus to pre-sacral nerve stimulation and to epinephrine, pituitrin and pilocarpine administration during certain sexual states in the anesthetized rabbit. Amer. J. Physiol. 111, 250—256 (1935).

Sawyer, W. H.: Neurohypophyseal secretions and their origin. In: Advances in neuroendocrinology. Urbana: Univ. Illinois Press 1963.

Schild, H. O.: The action of isoprenaline in the depolarized rat uterus. Brit. J. Pharmacol. 31, 578—592 (1967).

Schofield, B. M.: The innervation of the cervix and cornu uteri in the rabbit. J. Physiol. (Lond.) 117, 317—328 (1952).

— The influence of the ovarian hormones on myometrial behaviour in the intact rabbit. J. Physiol. (Lond.) 129, 289—304 (1955).

Setekleiv, J.: Uterine motility of the estrogenized rabbit. III. Response to hypogastric and splanchnic nerve stimulation. Acta physiol. scand. 62, 137—149 (1964).

Shore, P., Olin, J. S.: Identification and chemical assay of norepinephrine in brain and other tissues. J. Pharmacol. exp. Ther. 122, 295—300 (1958).

Silva, D. G.: The ultrastructure of the myometrium of the rat with special reference to the innervation. Anat. Rec. 158, 21—34 (1966).

Sjöberg, N. O.: The adrenergic transmitter of the female reproductive tract: distribution and functional changes. Acta. physiol. scand., Suppl. 305, 5—26 (1967).

— Increase in transmitter content of adrenergic nerves in the reproductive tract of female rabbits after oestrogen treatment. Acta endocr. (Kbh.) 57, 405—413 (1968).

SJÖSTRAND, N. O.: The adrenergic innervation of the vas deferens and the accessory male genital glands. Acta. physiol. scand. **65**, 9—77 (1965).

SOBOTTA, J.: Beiträge zur vergleichenden Anatomie und Entwicklungsgeschichte der Uterusmuskulatur. Arch. mikr. Anat. **38**, 52—100 (1891).

SPEDEN, R. S.: Electrical activity of single smooth muscle cells of the mesenteric artery produced by splanchnic nerve stimulation in the guinea-pig. Nature (Lond.) **202**, 193—194 (1964).

SPRATTO, G. R., MILLER, J. W.: The effect of various estrogens on the weight, catecholamine content and rate of contractions of rat uteri. J. Pharmacol. exp. Ther. **161**, 1—6 (1968a).

— — An investigation of the mechanism by which estradiol-17 β elevates the epinephrine content of the rat uterus. J. Pharmacol. exp. Ther. **161**, 7—13 (1968b).

STJÄRNE, L., LISHAJKO, F.: Comparison of spontaneous loss of catecholamines and ATP *in vitro* from isolated bovine adrenomedullary, vesicular gland, vas deferens and splenic nerve granules. J. Neurochem. **13**, 1213—1216 (1966).

STROUP, P. E.: Influence of epinephrine on uterine contractility. Amer. J. Obstet. Gynec. **84**, 595—601 (1962).

SULLIVAN, S. F., MARSHALL, J. M.: A quantitative evaluation of the effects of catecholamines on the human myometrium *in vitro*. Amer. J. Obstet. Gynec. (in press) (1969).

SWEDIN, G., BRUNDIN, J. O.: Distribution of noradrenaline in the genital organs of the female rat with a remark on dopamine in the cervix and vagina. Experientia (Basel) **24**, 1015—1016 (1968).

THEOBALD, G. W.: Nervous control of uterine activity. Clin. Obstet. Gynec. **11**, 15—33 (1968).

TOTHILL, ANNE: Investigation of adrenaline reversal in the rat uterus by the induction of resistance to isoprenaline. Brit. J. Pharmacol. **29**, 291—301 (1967).

TSAI, T. H., FLEMING, W. W.: The adrenotropic receptors of the cat uterus. J. Pharmacol. exp. Ther. **143**, 268—272 (1964).

VANOV, S., VOGT, M.: Catecholamine-containing structures in the hypogastric nerves of the dog. J. Physiol. (Lond.) **168**, 939—944 (1963).

VARAGIĆ, V.: An isolated hypogastric-nerve-uterus preparation, with observations on the hypogastric transmitter. J. Physiol. (Lond.) **132**, 92—99 (1956).

VOGT, M.: Transmitter released in the cat uterus by stimulation of the hypogastric nerves. J. Physiol. (Lond.) **179**, 163—171 (1965).

WANSBROUGH, H., NAKANISHI, H., WOOD, C.: Effect of epinephrine on human uterine activity *in vitro* and *in vivo*. Obstet. and Gynec. **30**, 779—789 (1967).

WILLEMS, J. L., SCHAEPDRYVER, A. F. DE: Adrenergic receptors in the oestradiol and allyl-oestrenol dominated rabbit uterus. Arch. int. Pharmacodyn. **161**, 269—274 (1966).

WURTMAN, R. J., AXELROD, J., KOPIN, I. J.: Uterine epinephrine and blood flow in pregnant and postparturient rats. Endocrinology **73**, 501—503 (1963).

— — POTTER, L. T.: The disposition of catecholamines in the rat uterus and the effect of drugs and hormones. J. Pharmacol. exp. Ther. **144**, 150—155 (1964).

— CHU, W., AXELROD, J.: Relation between the estrous cycle and the binding of catecholamines in the rat uterus. Nature (Lond.) **198**, 547—548 (1963).

Die Rolle des Carnitins im Intermediärstoffwechsel*

W.-D. THOMITZEK

Mit 2 Abbildungen

Inhaltsverzeichnis

Einleitung

Im Jahre 1905 wurde gleichzeitig von GULEWITSCH und KRIMBERG in Rußland und von KUTSCHER in Deutschland ein Stoff im Muskel entdeckt, der als Carnitin bzw. Novain bezeichnet wurde. Die Bezeichnung Carnitin setzte sich später durch. Erst 1927 klärten TOMITA und SENDJU die genaue Strukturformel: Carnitin ist das Trimethylbetain der γ-Amino-β-hydroxy-buttersäure (I).

$$CH_3-N^+-CH_2-CHOH \qquad CH_3-N^+-CH_2-CH_2OH \qquad CH_3-N-CH_2-CHOH$$

Carnitin (I) · · · · · · · · · Cholin (II) · · · · · · · · · · β-Methylcholin (III)

Nach ersten sporadischen Untersuchungen über Wirkungen von Carnitin im Organismus [8, 74] begann E. STRACK Anfang der dreißiger Jahre mit systematischen Untersuchungen über Carnitin und verschiedene seiner Deri-

* Aus dem Hauptlaboratorium des Stadtkrankenhauses, 7033 Leipzig, Friesenstr. 8 (Direktor: Prof. Dr. med. habil. D. LOHMANN).

Herrn Prof. Dr. Dr. Dr. h.c. Dr. h.c. E. STRACK zum 70. Geburtstag in Dankbarkeit.

vate. Ausgangspunkt war die strukturelle Verwandtschaft zum Cholin (II); denn Carnitin kann als substituiertes β-Methylcholin (III), das in Form seines Acetylesters Substrat der Acetylcholinesterase ist, aufgefaßt werden. Dabei fanden STRACK u. Mitarb., daß Acetylcarnitin im Gegensatz zum Acetylcholin eine verschwindend geringe (viele 10000—100000mal schwächere) muskarin- und nicotinartige Wirkung hat [116]. Erst nach Verschluß der COO⁻-Gruppe tritt eine Verstärkung dieser Wirkung auf, die aber noch viele Male schwächer ist als die von Acetylcholin [117]. Eine biologische Bedeutung schien solchen am Carboxyl substituierten Carnitinen nicht zuzukommen, da ein Umsatz durch entsprechende Fermente und auch das Vorkommen solcher Ester im Gewebe nicht nachweisbar waren [118].

Die in die gleiche Richtung zielenden Untersuchungen der belgischen Forscher BINON, CHARLIER, DALLEMAGNE, DELTOUR u. a. [40] mit zahlreichen Carnitinderivaten haben pharmakologische Bedeutung erlangt, ohne aber zur Klärung der physiologischen Funktion des Carnitins beizutragen. Zusammenhänge zwischen Carnitin und Cholin ließen sich nicht entdecken [39], ebensowenig kann Carnitin als Methylgruppendonator eingesetzt werden [46, 57, 140, weitere Lit. 107]. Allerdings soll es bei protein- und methioninarmer Diät einen Spareffekt für Methionin hervorrufen [73].

Als BENDER und ADAMS 1962 eine kritische Nachuntersuchung [6] zahlreicher angeblicher Wirkungen von Carnitin durchführten, konnten sie die meisten dieser Effekte nicht bestätigen und mußten feststellen, wie gering die Kenntnis von der physiologischen Rolle dieses Stoffes ist, der immerhin im Muskel in der hohen Konzentration von 3,5 % des Trockengewichts des eiweißfreien Extrakts vorkommt.

Entscheidende Fortschritte bei der Aufklärung der Funktion wurden dann seit Ende der fünfziger Jahre gemacht und die Bedeutung des Carnitins für den Fettsäurestoffwechsel kann als gesichert gelten, ohne daß noch andere Wirkungen für ausgeschlossen erklärt werden sollen.

Vier Faktoren haben die Erforschung der physiologischen Bedeutung des Carnitins entscheidend gefördert bzw. ermöglicht:

1. FRITZ beobachtete 1955 bzw. 1957, daß die Fettsäureoxydation in Leberhomogenaten durch Carnitin erheblich gesteigert wird [45, 46].

2. FRIEDMAN und FRAENKEL konnten 1955 nachweisen, daß in Taubenleberextrakten ein Enzym enthalten ist, das Carnitin reversibel acetylieren und deacetylieren kann [43].

3. STRACK und LORENZ veröffentlichten 1960 (verbessert 1966) eine einfache Synthese von reinem Carnitin und beschrieben u.a. eine rationelle Darstellung des natürlich vorkommenden (—)-Carnitins aus dem Razemat sowie die O-Acylderivate, obwohl damals das Vorkommen solcher Ester im Organismus höchstens zu vermuten war [120, 121].

4. Fritz u. Mitarb. entwickelten 1963 eine enzymatische Bestimmungs-methode für Carnitin [89] und ermöglichten damit, die unsicheren Angaben über den Gewebsgehalt an Carnitin, die mit chemischen [24, 44, 119] und biologischen [38, 119] Verfahren erzielt worden waren, zu präzisieren.

Vorkommen, Turnover und Biosynthese des Carnitins

Wie Tabelle 1 zeigt, hat die Muskulatur von den großen Stoffwechsel-organen den höchsten Gehalt an Carnitin. Das Fettgewebe weist nicht viel weniger auf. Dagegen enthalten Leber, Niere und Gehirn kleinere Mengen von

Tabelle 1. *Vorkommen von Carnitin und Acylcarnitinen*

Organ	Carnitin		Acetylcarnitin		Palmityl-carnitin %[a]
	µg/g trocken	Autoren	µg/g trocken	Autoren	
Nebennieren	~6 500	1, 28, 89, 90	—	—	—
Herzmuskel	400—1000	1, 28, 89, 90	350	89, 90, 102, 103	4(—23)
Fettgewebe	~500	89, 90	110	36, 89, 90	2(—12)
Skeletmuskel	250—550	1, 28, 89, 90	70—100	89, 90	—
Hoden	170—670	1, 28, 89, 90	110	89, 90	—
Cauda epididymus	14 000	91	750	91	—
Pankreas	350	28	—	—	—
Leber	150—210	1, 28, 36	30—60	36, 89, 90, 102	3(—16)
Gehirn	60—180	1, 28, 89, 90	30	89, 90	—
Niere	100—210	1, 28, 89, 90	120	89, 90	3(—8)
Darm	100—210	1, 28	—	—	—
Serum	≈ 10	89, 90	—	—	—

[a] % des Gesamtcarnitins [11, 12].

Carnitin. Auffallend hoch ist der Gehalt in den Nebennieren. Er soll in Mark und Rinde gleich groß sein [1]. Das ist allerdings nicht unwidersprochen ge-blieben [146]. Sehr viel Carnitin ist auch in den Hoden, insbesondere den Nebenhoden gefunden worden [89—91]. Erste Untersuchungen von Wolf und Berger [145] über den Carnitin-Turnover bei Ratten ergaben die sehr große Turnover-Zeit von rund 60 Tagen und einen Body-Pool von 35—40 mg/100 g Körpergewicht. Neuerdings konnte sowohl von Khairallah und Mehlman [72] als auch von Therriault und Mehlman [128] nach Ausschaltung von Fehler-quellen die Turnover-Zeit zu ca. 12 Tagen bestimmt werden. Bei Schwanger-schaft, Fasten und Cholinmangel beträgt sie sogar nur 5 Tage. Auch nach Kälteakklimatisation verkürzt sich die Turnover-Zeit auf die Hälfte [128].

Aus dem Body-Pool errechnet sich bei Annahme eines durchschnittlichen Wassergehalts des Organismus von 75 % ein Carnitingehalt von 1—1,5 mg/g Trockengewicht, also mehr als in Herz und Muskel vorliegen, die den höchsten Gehalt der großen Organe aufweisen. Daraus folgt, daß in einigen der kleinen Organe der Gehalt sehr viel höher sein muß. Bisher ist dies nur für die Nebennieren und Nebenhoden nachgewiesen worden.

Die tägliche Neusynthese des Carnitins in Ratten beträgt etwa 1 mg. Als Vorstufe gilt das γ-Butyrobetain, das zu Carnitin hydroxyliert wird [16, 79, 81—83]. Dieser Vorgang ist aber keine β-Oxydation, wie ursprünglich von LINNEWEH angenommen wurde [89], sondern erfolgt durch Oxydation mit Hilfe von Schwermetallen [83].

Die Methylgruppen des Carnitins stammen vom Methionin [15, 145] und Cholin [124, 125]. Bei Methionin- und Cholinmangeldiät sinkt der Carnitingehalt [124, vgl. auch 34]. Unklar ist noch die Herkunft des Kohlenstoffgerüstes, da nach übereinstimmenden Untersuchungen von BREMER [16] und LINDSTEDT [82] γ-Aminobuttersäure nicht die Vorstufe ist.

Carnitin und aktivierte Essigsäure

Die von FRIEDMAN und FRAENKEL [43] zuerst beschriebene reversible Acetylierung von Carnitin und Coenzym A (CoA) verläuft nach folgender Reaktion:

$$\text{Carnitin} + \text{Acetyl-CoA} \rightleftharpoons \text{O-Acetyl-Carnitin} + \text{CoA} \qquad (1)$$

In der Folgezeit wurden zahlreiche Acceptoren für den Acetylrest des Acetylcarnitins beschrieben. In vitro konnte die Übertragung auf Cholin unter Bildung von Acetylcholin [129—131], auf Oxalacetat unter Bildung von Citronensäure [51, 130] und auf Sulfanilamid [25, 43] nachgewiesen werden. Diesen Transacetylierungen ist gemeinsam, daß Coenzym A bei allen Reaktionen essentiell ist. Als Intermediärprodukt wird also stets Acetyl-CoA gebildet.

Die Wirksamkeit von Carnitin als Acetylacceptor zeigten THOMITZEK et al. [133] auch in vivo an der Hemmung der Acetylierung von Sulfanilamid sowie an der Senkung des Acetylcholinspiegels im Gehirn und Herzmuskel.

Acetylcarnitin — eine zweite Form aktivierter Essigsäure

Die Reaktion (1) kann in meßbarem Ausmaß nur dann reversibel sein, wenn die Esterbindung des O-Acetylcarnitins etwa den gleichen „Energieinhalt" hat wie Acetyl-CoA. Anhaltspunkte dafür gaben, nachdem die Erstbeschreibung durch FRIEDMAN und FRAENKEL [43] wenig beachtet worden war, etwa zur gleichen Zeit BREMER [17] und THOMITZEK [129—131]. Zweifel an dem energiereichen Charakter dieser Esterbindung wurden anfangs prinzipiell geäußert [65, 68, 112], konnten aber durch die Bestimmung der Gleichgewichtskonstanten endgültig widerlegt werden [51]. Danach beträgt die Hydrolyseenergie für Acetylcarnitin ($\Delta F^{o'}$ bei pH 7,0) —7,9 kcal, ist also nur wenig kleiner als die des Acetyl-CoA mit —8,2 kcal. Diese Ergebnisse wurden mit einem angereicherten Enzympräparat erhalten. Unterdessen ist das Ferment kristallisiert erhalten worden [30]. Es wird als Acetyl-CoA: Carnitin-O-Acetyltransferase (E. C. 2.3.1.7.) bezeichnet.

Hinsichtlich der sterischen Spezifität der Acetylübertragung läßt sich zeigen, daß das Derivat des natürlich vorkommenden linksdrehenden Carnitins wirksamer ist als das (+)-Isomere. Während BREMER [17, 18] aus seinen Versuchen mit (—)- und (±)-Carnitin auf eine Wirkungslosigkeit des (+)-Isomeren schloß, fand THOMITZEK [129, 130], daß reines (+)-Acetylcarnitin eine zwar stark abgeschwächte, aber meßbare Wirkung bei der Transacetylierung zeigt. Von weiteren Versuchen in dieser Richtung seien nur die von FRITZ und SCHULTZ [55] erwähnt, die eine Hemmwirkung des (+)-Carnitins auf das gereinigte Enzym beobachteten. Bei vielen Versuchen ist aber wohl nicht genügend berücksichtigt worden, daß z.B. (—)-Acetylcarnitin nicht nur Substrat, sondern auch Hemmstoff der Transferase ist [55]. Die sterische Spezifität ist also ausgeprägt, aber nicht absolut. Die Prüfung von Strukturabwandlungen des Carnitins auf den Acetylübertragungsmechanismus ergab, daß die OH-Gruppe essentiell ist. γ-Butyrobetain ist wirkungslos bzw. ein Hemmstoff. Hingegen ist der quaternäre Stickstoff von geringerer Bedeutung: Das um eine Methylgruppe ärmere „Norcarnitin" (γ-Dimethylamino-β-hydroxybuttersäure) zeigt zwar eine geringere Affinität (Michaeliskonstante 1,2 mM gegenüber Carnitin 0,4 mM) [55, 131], mit hohen Konzentrationen wird aber eine fast gleich große Maximalgeschwindigkeit erreicht. Analog wie bei anderen Fermenten mit einem anionischen Bindungsort (z.B. Acetylcholinesterase) ist also die protonisierte Dimethylaminogruppe auch zur Bindung befähigt. Wird jedoch die Demethylierung bis zur γ-Amino-β-hydroxybuttersäure fortgesetzt, so erlischt die Wirkung völlig [49, 130, 131]. Die freie Carboxylgruppe (COO^-) ist ebenfalls bedeutsam. Wird sie durch die Nitril- (CN), Ester-(COO—R) oder Amid-($CO—NH_2$)-Gruppe ersetzt, so tritt Verlust [46] bzw. starke Einbuße [129, 131] der Wirkung ein.

Die Übertragung kurz- und langkettiger Fettsäuren

Prüft man die Spezifität der Acetyltransferase hinsichtlich der Kettenlänge der zu übertragenden Fettsäure, so findet man bei optimalem pH die größte Aktivität mit Essigsäure. Bereits Propionsäure und Buttersäure zeigen deutlich verringerte Maximalgeschwindigkeiten [32]. Frühere gegenteilige Angaben [51] können als widerlegt gelten, da sie nur bei einer einzigen Substratkonzentration durchgeführt wurden. Exakte Schlüsse auf die Affinität zum Ferment lassen sich daraus nicht ziehen. Mit zunehmender Kettenlänge des

Fettsäurerestes nimmt die Aktivität immer mehr ab und mit Myristinsäure ist sie Null. Carnitin fördert jedoch in Gewebspräparationen auch die Oxydation langkettiger Fettsäuren, so daß ein weiteres Enzym für langkettige Fettsäuren zu vermuten war, das durch FRITZ und YUE [50, 52] eindeutig nachgewiesen werden konnte. Etwa zur gleichen Zeit fand BREMER die Bildung von Palmitylcarnitin im Gewebe [19]. Exakte Aktivitätsbestimmungen gelangen jedoch erst NORUM, der die störende Palmityl-CoA-Hydrolase abtrennen konnte [98]. Das Ferment ist als Acyl-CoA: Carnitin O-Acyltransferase zu bezeichnen und wird im folgenden kurz „Palmityltransferase" genannt. Die Aktivität sinkt mit fallender Kettenlänge (Tabelle 2) [98]. Allerdings gilt für Schlüsse aus diesen Werten wieder die obenerwähnte Einschränkung bei Messungen mit nur einer einzigen Substratkonzentration.

Table 2. *Die Abhängigkeit der Reaktionsgeschwindigkeit der Carnitin-Acyltransferasen von der Kettenlänge der Acyl-CoA-Derivate*

Fettsäure-Rest	C-Atome	Acetyl-transferase[a] V_{Max} %	Palmityl-transferase[b] v %
Essigsäure	2	100	15
Propionsäure	3	77	—
Buttersäure	4	41	6
Caprylsäure	8	8	32
Caprinsäure	10	4	50
Laurinsäure	12	0,1	65
Myristinsäure	14	0	88
Palmitinsäure	16	0	100

[a] V_{Max} bei pH 7,8 und 30° C graphisch ermittelt und auf Essigsäure ($=100\%$) bezogen (CHASE[32]).

[b] v bei pH 7,5 und 0,4 µMol D,L-Carnitin, 0,18 µMol CoA in 1,5 ml Endvolumen ermittelt und auf Palmityl-CoA ($=100\%$) bezogen (NORUM[98]).

Die nun naheliegende Frage nach dem Gehalt der Gewebe an Acylcarnitinen versuchten zuerst PEARSON und TUBBS zu beantworten [102]. Neueste Ergebnisse sind in der Tabelle 1 aufgeführt. Das Acetylcarnitin beträgt etwa 20%, das Palmitylcarnitin etwa 2% des freien Carnitins [11, 12, 36, 90, 91]. Bei fettreicher Kost, im Hunger oder bei Diabetes ist der Palmitylcarnitingehalt erheblich vermehrt (s. Tabelle 1, Werte in Klammern).

Zahlreiche Autoren fanden Carnitin in lipidgebundener Form. MEHLMAN und WOLF beschrieben die Existenz von Phosphatiden mit gebundenem Carnitin als N+-Base (Phosphatidylcarnitin) [93, 94]. Nach neueren Ergebnissen kann aber als sicher gelten, daß es sich hierbei um an die Phosphatide assoziativ gebundene langkettige Acylcarnitine gehandelt hat [42, 54, 114]. Carnitin in echter chemischer Bindung (Esterbindung) kommt nicht oder höchstens in Spuren in Phosphatiden vor. Der polare Charakter des Carnitins und die hydrophobe Fettsäureseitenkette machen relativ feste Anlagerungsverbindungen wahrscheinlich. Palmitylcarnitin zeigt ohnehin eine erhebliche Neigung zu Micellenbildung (unveröffentlichte eigene Versuche).

Die Verteilung der Carnitin-Acetyltransferase in verschiedenen Organen ist in der Tabelle 3 aufgeführt. Die zuverlässigsten Werte scheinen die von McCaman et al. [92] und Norum und Bremer [100] zu sein, die 5- bis 20mal höher liegen als die anderer Autoren [5, 87]. Hohe Aktivitäten besitzen besonders solche Organe, von denen bekannt ist, daß sie einen erheblichen Teil ihres Energiebedarfs durch Oxydation von Fettsäuren decken. Die Palmityltransferase-Aktivität ist nach Fasten, fettreicher Diät und beim Diabetes erheblich gesteigert [99], also bei Stoffwechselsituationen mit gesteigerter Fettsäureoxydation.

Tabelle 3. *Aktivitäten der Carnitin-Acetyltransferase*

	Organ		Mitochondrien
	Mol/min · g trocken[a]	Mol/h · g feucht[b]	Mol/min · g Mitoch.-Protein[a]
Herz	170	1600 (~107)	440
Fettgewebe	100	—	—
Skeletmuskel	25	390 (~ 26)	410
Hoden	145	905 (~ 60)	—
Niere	30	413 (~ 27)	109
Gehirn	15	110 (~ 7,3)	12
Nebennieren	—	105 (~ 7)	—
Leber	—	70 (~ 5)	5

[a] Nach Marquis [90].
[b] Nach McCaman et al. [92]. In Klammern auf Mol/min · g trocken umgerechnete Werte unter Annahme eines Wassergehalts von 75%.

Die intracelluläre Verteilung des Carnitins und der Transacylasen

Wie die Tabelle 4 zeigt, befinden sich 85% des freien Carnitins im Cytoplasma (100000 × g Überstand), während Kern und Mitochondrien die restlichen 15% enthalten. Die Mikrosomen sind offenbar frei von Carnitin [90].

Tabelle 4. *Freies Carnitin in cellulären Subfraktionen.* (Nach Marquis und Fritz [90])

			µMol/g trocken	%
1 000 × g Sediment	≙	Kerne	0,36	8,1
25 000 × g Sediment	≙	Mitochondrien	0,25	5,6
100 000 × g Sediment	≙	„Mikrosomen"	0,0	0
100 000 × g Überstand	≙	„Cytoplasma"	3,83	86

Die Fraktionierung erfolgte in 0,25 M Saccharose. Organ: Rattenherz.

Die beiden Transacylasen sind jedoch fast ausschließlich mitochondrial lokalisiert [5, 90, 92, 100] (Tabelle 5). Die mitochondriale Lokalisation scheint für die Palmityltransferase sogar so typisch zu sein, daß sie als Markierungsenzym bei Zellfraktionierungen vorgeschlagen wurde [100]. Kürzlich ist berichtet worden, daß das Ferment an der Innenmembran des Mitochondrion fixiert sein soll [100], während es nach ALLMANN u. Mitarb. [2, 3] an der Außenmembran sitzen soll. Die Lokalisation der Transacylasen in den Mitochondrienmembranen ist für die Rolle des Carnitins als transmembranaler Fettsäure-Carrier von Bedeutung.

Tabelle 5. *Intracelluläre Verteilung der Acetyl- und Palmityltransferase*

Zellfraktionen von	Acetyltransferase		Palmityl-transferase
	Leber[a] %	Gehirn[b] %	Leber[a] %
Kerne	11,6	6,5	15,5
schwere Mitochondrien	56,7	73	65,3
leichte Mitochondrien	5,3	—	1,4
Mikrosomen	18,2	20	8,0
„Cytoplasma"	3,4	0,5	3,5

[a] Nach NORUM und BREMER [100].
[b] Nach McCAMAN et al. [92].

Carnitin und Fettsäurestoffwechsel

Seit langem wissen wir auf Grund von Versuchen an Gewebsschnitten [59, 141] sowie isolierten bzw. perfundierten Organen [z.B. 47, 64, 144], daß viele Organe Fettsäuren verbrennen können. Dafür sprach auch das Verhalten des respiratorischen Quotienten [9, 35, 75] und das Vorliegen von arterio-venösen Differenzen des Fettsäurespiegels [4, 60]. Um so erstaunlicher war die vielfach beobachtete Tatsache, daß in vitro mit Mitochondrien der meisten Organe, die in vivo eine lebhafte Fettsäureoxydation zeigen, nur ein geringer Umsatz der Fettsäuren nachweisbar war [95, 97, 104]. Eine Ausnahme machen lediglich die Lebermitochondrien, die langkettige Fettsäuren in merklichem Umfang oxydieren [70, 71, 78]. Paradoxerweise wurde die fördernde Wirkung des Carnitins aber bei Untersuchungen über die Fettsäureoxydation in Leberhomogenaten entdeckt [45]. An der weiteren Erforschung dieses Mechanismus sind u.a. neben FRITZ u. Mitarb. besonders die Arbeitskreise von BREMER in Oslo und KLINGENBERG in Marburg beteiligt gewesen.

Um die Rolle des Carnitins bei der Fettsäureoxydation verstehen zu können, müssen wir uns die bekannten Tatsachen des Fettsäureabbaus vergegenwärtigen.

Die Fettsäureoxydation wird durch die Aktivierung der Fettsäure mit Hilfe der Acyl-CoA-Synthetase eingeleitet:

$$R-COOH + CoA + ATP \rightarrow R-CO-CoA + AMP + \text{Pyrophosphat}$$

Die aktivierte Fettsäure wird dann nach dem Schema der β-Oxydation zu Acetyl-CoA, dieses im Citronensäurecyclus zu CO_2 abgebaut [zusammenfassende Darstellung s. 88]. Während die Oxidation der aktivierten Fettsäuren in den Mitochondrien erfolgt, läuft die Aktivierung offenbar überwiegend extramitochondrial ab [zur Aktivierung mit GTP vgl. 109]. Es muß also ein Transport der aktivierten Säure von dem extramitochondrialen Bildungsort in das Mitochondrion erfolgen. Die Acyl-CoA-Verbindungen sind dafür schlecht geeignet, da Nucleotide die Mitochondrienmembran nur langsam penetrieren können [87, 105]. Aus seinen Versuchen mit Carnitin konnte Fritz bereits

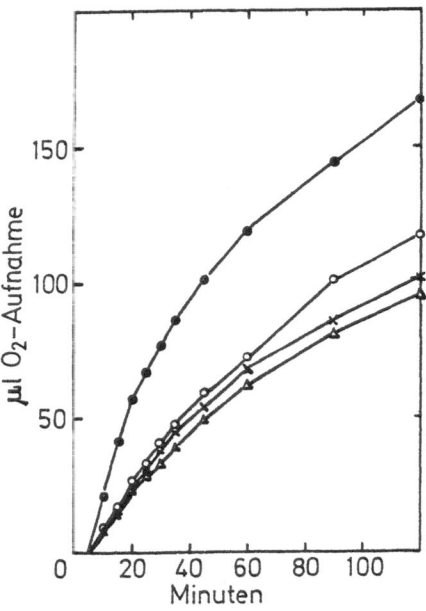

Abb. 1. Einfluß von Carnitin und -derivaten auf die Atmungsrate von Herzmuskelmitochondrien. △—△ Palmitinsäure (0,1 mM) oder kein zugesetztes Substrat; ×—× Palmitinsäure und D,L-Carnitin (0,5 mM); ○—○ Palmityl-CoA (0,1 mM); ●—● D,L-Palmitylcarnitin (0,1 mM). Versuchsbedingungen: 5,6 mg Mitochondrienprotein in 2,5 ml Inkubationsmedium von 0,25 M Saccharose, 1 mM Na-EDTA, 1 mM MgCl, 60 mM KCl, 3,33 mM AMP, 6 mM Natriumphosphatpuffer pH 7,4; 21 mg Rinderserum-albumin, 1 mM Succinat. (Nach [52])

1959 den Schluß ziehen, daß Carnitin am Fettsäuretransport angreifen muß [48]. Heute kann als gesichert gelten, daß Acylcarnitin schneller als Acyl-CoA oder Fettsäuren mit allen zur Aktivierung benötigten Cofaktoren oxydiert wird (s. Abb. 1). Die schnellere Oxydation des Acylcarnitins gegenüber dem Acyl-CoA ist nur durch einen besseren Zugang zum Oxidasesystem zu erklären, denn zugesetzte Acylcarnitine umgehen die Aktivierungsreaktion, die bei der Oxydation freier Fettsäuren notwendig ist, da sie bereits aktivierte Acyle wie Acyl-CoA darstellen. Oxydationsraten mit Acylcarnitinen stellen somit die Maximal-Kapazität der mitochondrialen Fettsäureoxydation dar [10] und sind für die vergleichende Physiologie bedeutungsvoll. In der Zelle werden die Acylcarnitine nur über die Acyl-CoA-Verbindungen gebildet. Eine direkte Bildung im Sinne einer Carnitinligase scheint nicht möglich zu sein. Auch eine Transferase zwischen kurz- und langkettigen Acylcarnitinen, die aus einigen bisher unerklärten experimentellen Befunden als möglich postuliert wurde [10], konnte trotz intensiver Suche nicht nachgewiesen werden [61, 139].

Die Rolle des Carnitins als Carrier für Fettsäuren ähnelt anderen transmembranalen Transportmechanismen, mit deren Hilfe in der Zelle „Gruppen" durch die Zellmembran transportiert werden, da eine einfache Diffusion entweder auf Grund der Löslichkeitsverhältnisse oder anderer Ursachen für die Bedürfnisse der Zelle nicht ausreichend schnell ist. Als Beispiele seien die zahlreichen „Shuttles" erwähnt, die z. B. den Transport von Wasserstoff, der bei den Dehydrogenasereaktionen in Form von NADH anfällt, durch Membranen ermöglichen. Ein solcher Shuttle-Mechanismus für Fettsäuren läuft mit Hilfe des Carnitins ab. Zweifel an der Carrieraufgabe des Carnitins sind geäußert worden, da die transferierenden Enzyme offenbar ausschließlich, sicher aber überwiegend mitochondrial lokalisiert sind. Die Wirksamkeit der bekannten Shuttles ist jedoch mit zwei verschiedenen Enzymen, einem mitochondrialen und einem cytoplasmatischen, verknüpft. Zur Erklärung ist anzuführen, daß die Carnitin-Acyltransferasen in der Mitochondrienmembran sitzen und von beiden Seiten zugängig sein könnten und damit die Rolle bei der Translokation voll erfüllen könnten. Neuerdings ist es mit spezifischen Hemmstoffen gelungen, das System der mitochondrialen Fettsäureoxydation funktionell weiter zu differenzieren. Danach existieren zwei Barrieren, die möglicherweise in der Außen- und Innenmembran des Mitochondrions ihr morphologisches Äquivalent haben. Nur Acylcarnitin ist in der Lage, beide Barrieren zu passieren und als Mittler zwischen den in beiden Kompartimenten vorhandenen, nicht frei austauschbaren CoA-Pools zu fungieren [29, 61] (Abb. 2). Damit kommt dem Carnitin nicht nur beim Fettsäuretransport zwischen Cytoplasma und Mitochondrion eine wesentliche Rolle zu, sondern auch zwischen verschiedenen intramitochondrialen Kompartimenten.

Das Vorhandensein von Carnitin in Geweben ohne langkettige Fettsäureoxydation [33] weist auf seine Rolle bei der Transacetylierung hin.

Carnitin wirkt auch am Transport von Fettsäuren in der Richtung vom Mitochondrion zum Cytoplasma mit. Wenn Überfluß an Nährstoffen besteht, werden intramitochondrial größere Mengen an Acetyl-CoA gebildet, als oxydiert werden können. Diese werden im allgemeinen extramitochondrial zur Fettsäuresynthese und Acetylierungsreaktionen genutzt. Allerdings zeigen Experimente das Überwiegen des Citrat-Citrat-Cleavage-Systems für die Translokalisation des aktivierten Acetats zur Fettsäuresynthese [87, 115], obwohl Carnitin dabei auch mitwirken kann [22—24, 53]. Dagegen spielt Carnitin offenbar eine erhebliche Rolle bei anderen extramitochondrialen Acetylierungsreaktionen [25].

An der Rolle des Carnitins als Fettsäurecarrier kann heute kein Zweifel mehr bestehen. Dafür sprechen neben den angeführten Befunden der hohe

Turnover des Acetylcarnitins mit einer Halbwertszeit von nur wenigen Minuten [36], die Tatsache, daß der Gehalt an Carnitin-Palmityltransferase so hoch ist, daß er die Oxydationsrate langkettiger Fettsäuren voll erklären kann [100], sowie die Hemmbarkeit von Stoffwechselvorgängen, die durch den Membrantransport begrenzt sind, durch Derivate des (+)-Carnitins [53].

Bei der Betrachtung der Rolle des Carnitins für den Stoffwechsel von Acyl- bzw. Acetyl-CoA dürfen Situationen nicht vergessen werden, in denen ein starker Überschuß von Acetyl-CoA zur Bildung der Acetessigsäure und der anderen Ketokörper führt.

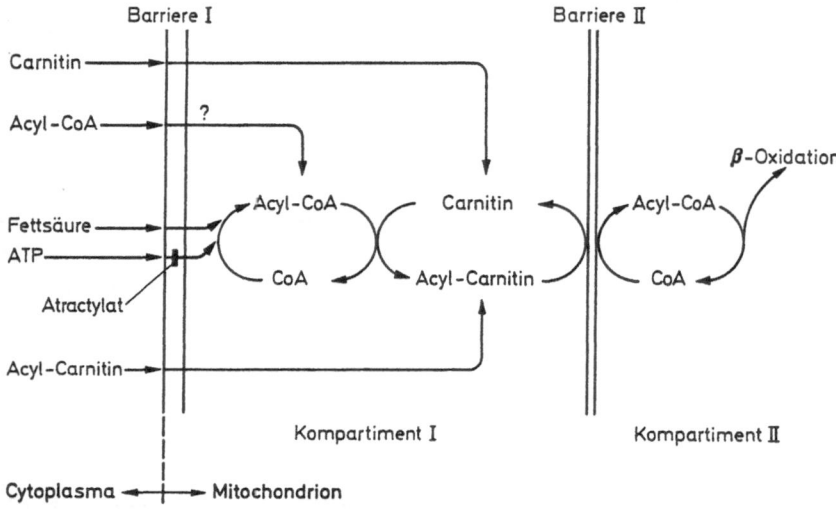

Abb. 2. Die Vermittlerrolle des Carnitin bei dem Transport von Fettsäuren zwischen zwei mitochondrialen Kompartimenten (in Anlehnung an [29]). Im Kompartiment I läßt sich die ATP-abhängige Fettsäureaktivierung durch Atractylat, einen Hemmstoff der ATP-Translokation, hemmen

Carnitin und Ketokörper

Bereits Fritz zeigte bei seinen ersten Versuchen über die Wirkung des Carnitins auf die Fettsäureoxidation, daß im Verhältnis viel Ketokörper gebildet werden [45]. Bremer [17] und Strack u. Mitarb. [76, 122] bestätigten, daß aus kurz- und langkettigen Acylcarnitinen Acetacetat gebildet wird. Bode und Klingenberg [10] präzisierten diese Ergebnisse und zeigten u.a., daß Citronensäure-Cyclus-Intermediate zur Verwertung des Acetyl-CoA bzw. Acetylcarnitins notwendig sind. Damit wurden die vorher von Bressler und Katz [22, 23] erhaltenen Befunde über die Abhängigkeit der Ketokörperbildung unter Carnitin vom vorherigen Ernährungszustand gut erklärt. Während in Leberhomogenaten gefütterter Tiere die Ketokörperbildung durch Carnitin um etwa 30% gehemmt war, steigerte Carnitin die erhöhte Ketogenese in Leberpräparationen fastender Tiere auf das Doppelte.

Diesen Versuchen in vitro stehen die Ergebnisse in vivo gegenüber. Bei normalen Tieren fanden Strack et al. [122] erhöhte Ketokörperwerte nach

Carnitingaben im Blut. Dagegen konnte Carnitin die durch Fettfütterung oder Fasten hervorgerufene Acidose und Ketose bei Versuchstieren prompt und vollständig beseitigen [27, 108].

Diese Befunde sind aus der Kenntnis der Ketogenese als Mißverhältnis zwischen Kohlenhydrat- und Fettstoffwechsel heute plausibel zu erklären. Die Kenntnis der Rolle des Carnitins in diesem normalerweise gut abgestimmten Mechanismus kann neue therapeutische Möglichkeiten für die Klinik eröffnen.

Carnitin und Stoffwechselregulation

Wechselwirkungen zwischen Kohlenhydrat- und Fettstoffwechsel am Gesamtorganismus sind wohlbekannt und auch auf cellulärer bzw. subcellulärer Basis liegen seit Jahren zahlreiche Befunde vor, die zeigen, daß Glucose die Oxydation langkettiger Fettsäuren hemmt [z.B. 86], während die Fettsäuren die Oxydation der Glucose bremsen [z.B. 37]. Das gemeinsame Glied beider Stoffwechselwege ist das Acetyl-CoA, dessen Weiterverwertung auf drei Hauptwegen erfolgt. Es wird

1. im Citronensäurecyclus zu CO_2 und H_2O oxidiert,
2. zu langkettigen Fettsäuren resynthetisiert und
3. zu Ketokörpern kondensiert.

Über die Steuerung der Verteilung auf diese Wege wissen wir erst seit kurzer Zeit Näheres. Die langkettigen Fettsäure-CoA-Thioester scheinen eine dominierende Rolle dabei zu spielen. Sie sind in der Leber bei reichlichem Fettsäureangebot durch fettreiche Nahrung oder durch erhöhte Fettsäuremobilisation im Hunger und bei Diabetes auf das Vielfache erhöht [13, 136, 137, 143]. Diese Acylthioester greifen an vielen Fermentreaktionen an, bei denen sie als Substrate oder Reaktionsprodukte gar nicht in Erscheinung treten. Das Vorkommen solcher Effekte wird als Allosterie bezeichnet [96] und hat die Kenntnis und Erforschung der Stoffwechselregulation in hervorragender Weise gefördert. Die Acyl-CoA-Thioester entfalten solche allosterischen Hemmeffekte u.a. auf die Citratsynthase (Citronensäurebildung aus Oxalacetat und Acetyl-CoA) [135, 142] und auf die Acetyl-CoA-Carboxylase, die die Fettsäuresynthese aus Acetyl-CoA, CO_2 und ATP einleitet [14]. Durch den Ausfall dieser beiden Wege wird das im Stoffwechsel anfallende Acetyl-CoA zur Ketokörperbildung umgeleitet. Trotz gewisser unspezifischer Hemmwirkungen auf viele Enzyme [127] kommt den langkettigen Acyl-CoA-Thioestern sicher eine wichtige Regulationsfunktion in der Verwertung des Acetyl-CoA zu.

Es ist bekannt und aus dem Mechanismus der Ketokörperbildung zu verstehen, daß kleine Änderungen des Acetyl-CoA-Spiegels zu großen der Ketokörper führen [58, 61, 113, 137]. Die Umschaltung der Palmitylcarnitinoxydation von der Acetacetatbildung auf den Citronensäurecyclus geht z.B. nur mit einer Änderung des Acetyl-CoA-Spiegels um 20 % einher [61]. Neuerdings mehren sich allerdings die Befunde, die weniger den Spiegel an Acetyl-CoA als die Verfügbarkeit von freiem CoA [36, 111] bzw. das Verhältnis Acetyl-CoA/freies CoA [66, 67] als bestimmenden Faktor ansehen. Alle Faktoren, die diesen Quotienten vergrößern, führen zu einer vermehrten Ketogenese.

Carnitin steht aber, wie wir oben sahen, über wirksame Fermente mit den CoA-Thioestern in Verbindung, so daß Veränderungen der Carnitinester analog den Acyl-CoA-Verbindungen zu erwarten sind. Tatsächlich konnte Bøhmer [11, 12] unlängst zeigen, daß die langkettigen Acylcarnitine die Veränderungen der CoA-Thioester mitmachen, also wie diese bei Hunger, Diabetes und oraler Fettzufuhr erhöht sind (Tabelle 1, Werte in Klammern). Es besteht Grund zu der Annahme, daß die Carnitin-Acyltransferasen normalerweise ausreichende Aktivität besitzen, um eine Gleichgewichtseinstellung zwischen Carnitin und CoA zu gewährleisten [21, 36]. Damit erhält das Carnitin eine eminente Bedeutung für den Kohlenhydrat- und Fettsäurestoffwechsel, obwohl es nicht im Zuge der Abbaureaktionen, sondern nur im Nebenschluß liegt. Die Fähigkeit des Carnitins zur Penetration der Zellmembran eröffnet Perspektiven für die therapeutische Beeinflussung des Spiegels an verestertem und freiem CoA. Damit kann eine wirksame Unterstützung z. B. der Acidosetherapie bei Stoffwechselentgleisungen erwartet werden. Sporadische experimentelle Untersuchungen darüber wurden bereits erwähnt. Im Lichte unserer heutigen Kenntnisse ergeben sich jedoch einige neue Gesichtspunkte, die eine qualitative und quantitative Analyse rechtfertigen.

Nach Carnitinzufuhr ist durch Übertragung der aktivierten Fettsäuren von CoA auf Carnitin eine Senkung des Quotienten Acyl-CoA/freies CoA zu erwarten und damit eine Enthemmung der Citratsynthase und der Acetyl-CoA-Carboxylase. Das muß zu einer verminderten Ketokörperbildung führen. Außerdem wird die Oxydation der Ketokörper durch den vermehrten Umsatz im Citronensäurecyclus stimuliert [66, 67, 109].

Carnitin ist über den Mechanismus einer Freisetzung von CoA auch in der Lage, die Acetacetat-Oxidation in Nierenmitochondrien und Herzsarkosomen zu steigern [67].

Das Angebot an freiem CoA bestimmt auch das Ausmaß der Gluconeogenese mit. Wie erwähnt, hemmen Fettsäuren die Oxydation von Pyruvat [37, 101], und der Pyruvatspiegel und -umsatz sind bei vermehrtem Fettsäureangebot erhöht [106]. Das ist die Folge der verringerten Pyruvatdecarboxylierung [63]. Diese (als irreversibler Initialschnitt des oxydativen Pyruvatabbaus) bestimmt die Weiterverwertung und ist vom freien CoA im Mitochondrion abhängig [20]. Wird wenig Pyruvat decarboxyliert, so kann entsprechend mehr zu Oxalacetat (bzw. Malat) carboxyliert werden [41]. Acetyl-CoA wirkt dabei zudem als allosterischer Aktivator [69, 138]. Die resultierende vermehrte Gluconeogenese ist also ebenfalls Folge des Mangels an freiem CoA. Auch hier kann das Carnitin angreifen.

Diese Vorstellungen widersprechen nicht den Befunden von BENMILOUD und FREINKEL an Leberschnitten von Kaninchen [7], in denen durch Carnitin ein vermehrter Einbau von Alanin oder Pyruvat in Glucose bewirkt wurde; denn hier handelte es sich um Tiere in der Phase der Umschaltung von Kohlenhydrat- auf Fettsäurestoffwechsel. In diesem Zustand kann die Acylcarrierfunktion des Carnitins ein vermehrtes Acylangebot hervorrufen.

Nicht alle bekannten Nebenreaktionen der aktivierten Acyle lassen sich hier besprechen. Hinweise über eine intramitochondriale Kompartimentierung des Coenzyms A sind bekannt [29, 58]. In diesem Sinne oder durch andere noch unbekannte Mechanismen muß z. B. die erstaunliche Beobachtung gewertet werden, daß Acetylcarnitin an Mitochondrien nicht in der Lage ist, die $^{14}CO_2$-Bildung aus (U-^{14}C-Palmityl) Palmitylcarnitin zu senken, während der gleiche Effekt mit 2-^{14}C-Pyruvat prompt eintritt [20].

Neuerdings wissen wir auch, daß der Gehalt an freiem Carnitin bei Ketose in Leber und Muskel gleichermaßen auf ca. 50% vermindert ist, während erstaunlicherweise Acetylcarnitin nur in der Leber erhöht, im Muskel aber gesenkt ist. Auch der Acetylturnover ist in der Leber gesteigert, im Muskel vermindert. Ein ähnlich divergentes Verhalten besteht im Redox-Status der beiden Organe [36]. Um so bemerkenswerter ist die Übereinstimmung in der Abnahme an freiem Carnitin in beiden Organen. Eine Erhöhung durch therapeutische Zufuhr bietet sich daher geradezu an.

Pharmakokinetik des Carnitins

Die akute Toxizität des Carnitins ist erstaunlich gering. Die LD_{50} acuta beträgt bei der Maus 13,5 mg/g [110]. Demzufolge liegen die bisher therapeutisch angewendeten Dosen (beim Menschen per os 1—2 g/die, in Tierversuchen 20—300 mg/kg je 24 Std i.v.) im Bereich um 1% dieser toxischen Menge.

Die Eliminationskinetik zeigt, daß Carnitin schnell renal ausgeschieden wird. So sind bereits nach wenigen Stunden etwa 30% einer Stoßdosis (1 bis 2 mg/kg i.v.) im Urin ausgeschieden [147]. Nach dem ersten Tag sinkt die Ausscheidungsgeschwindigkeit jedoch sehr stark ab [80]. Das scheint daran zu liegen, daß Carnitin in den Geweben fixiert wird. 7 Std nach einer Injektion von 2 mg/kg i.v. bei Hunden ist die Konzentration (auf Zellwasser berechnet) in vielen Geweben höher als im Plasma [147], so im Herzen 10fach, in der Skeletmuskulatur 15—20fach, in der Leber 30fach und in der Niere 30—70fach. Lediglich im Gehirn ist sie nur $^1/_{16}$. Daraus auf einen aktiven Akkumulationsprozeß zu schließen [147], scheint etwas zweifelhaft, da man leicht nachrechnen kann, daß nach der Injektion die Plasmakonzentration etwa 100mal größer gewesen sein muß als nach 7 Std. Das Eindringen des Carnitins in die Gewebe kann also zu diesem Zeitpunkt allein durch den vorliegenden Konzentrationsgradienten erfolgt sein. Möglicherweise ist auch die Bezugsgröße

„celluläres Wasser" nicht geeignet, da eine Fixation des polaren Carnitinmoleküls an Membranen leicht denkbar ist und für Erythrocytenmembranen auch nachgewiesen wurde [132].

Betrachtet man die Mengenverteilung des applizierten Carnitins, so findet man auf Grund des großen Anteils der Muskulatur an der Körpermasse den überwiegenden Anteil der injizierten Dosis in der Muskulatur, 7 Std nach der Injektion (2 mg/kg) sind 55 % der Dosis in der Muskulatur enthalten, 3 % in der Leber, 1 % in den Nieren und weniger als 1 % in anderen Organen, während im Urin innerhalb der 7 Std 35 % ausgeschieden wurden [147]. Nach 7 Tagen ist die Anreicherung (Aktivität/g) in den einzelnen Organen etwas verändert, da das Herz das Carnitin offenbar besonders festhält (Werte von [80]). Der prozentuale Anteil an der Gesamtdosis ist für die Skeletmuskulatur jedoch immer noch bei weitem am größten (ca. 20 % der applizierten Dosis).

Bei oraler Gabe ist zu bedenken, daß Carnitin bei der Leberpassage in wechselndem Ausmaß acyliert werden kann und daß damit weniger freies Carnitin in die anderen Organe gelangen dürfte.

Bei Abschätzung der notwendigen effektiven Dosis sind folgende Gesichtspunkte zu berücksichtigen: erstens die organspezifische Anreicherung des Carnitins, zweitens die Größe des Pools des Carnitins und seiner Acyle in den einzelnen Organen und drittens der Turnover derselben. Die sehr kurze Halbwertszeit des Acetylturnovers in der Leber von nur 2—3 min [36] läßt Bedenken gegen eine wirksame Carnitintherapie entstehen. Jedoch beträgt die Halbwertszeit in der Muskulatur, die mehr als 50 % des Carnitins aufnimmt, etwa 15 min. Bei einem Gehalt an freiem Carnitin von 50—100 mg/kg Muskel im ketotischen Zustand enthält die Muskulatur des Menschen also 1—2 g! Bereits die Zufuhr von 1 g muß also zu einer spürbaren Änderung des intracellulären Carnitingehalts führen und damit auch den Spiegel an freiem Coenzym A erhöhen. Noch fühlbarer muß sich jedoch dieser Effekt bei den in viel geringerer Konzentration vorhandenen Acylcarnitinen bzw. Acyl-CoA-Thioestern auswirken. Eine ins einzelne gehende Analyse scheitert an folgenden ungeklärten Punkten: langkettiger Acyl-CoA- (und Carnitin-) Turnover, Zugängigkeit der einzelnen CoA-Kompartimente für Carnitin bzw. ihre Ausstattung mit der Transferase, Acceptorstärke des Carnitins für kurz- und langkettige Acyle, was nicht nur eine Frage der Fermentaffinität (Michaeliskonstante), sondern auch der cellulären Organisation (Kompartimente) sein muß u.a.

Des weiteren läßt sich die therapeutische Wirkung des Carnitins nicht im einzelnen voraussehen, da auf Veränderungen des Quotienten Acyl-CoA/freies CoA eine Vielfalt von Regulationen einsetzt. Da bisher eine Abnahme des Gesamtcarnitingehalts bei Stoffwechselstörungen noch nicht beschrieben wurde, ist sehr wahrscheinlich, daß der Carnitineffekt aufhört, sobald das Carnitin entsprechend dem vorliegenden Acyl-CoA/CoA-Quotienten acyliert ist. Damit ist vorauszusehen, daß das Carnitin nur für begrenzte Zeit wirksam sein wird. Es scheint also besonders geeignet bei akuten Entgleisungen, etwa bei hochgradiger diabetischer Acidose oder im Koma, am besten in Form langdauernder Infusionen. Mehrere Gramm Carnitin sind bisher ohne Nebenwirkungen gegeben worden [123]. Um den Effekt des Carnitins zu verbessern und zu verlängern, empfiehlt es sich, den Abfluß der aktivierten Fettsäuren zu erleichtern, wofür u. a. die Verwertung bei der Lipidsynthese in Frage kommt. Dem Mangel an α-Glycerophosphat, der beim Diabetes u. a. die Fettsynthese begrenzen kann, ist aber mit Fructose oder Sorbit abzuhelfen, die beide insulinunabhängig verwertet werden und selbst bereits kräftig antiketogen wirken.

Abschließend soll nicht unerwähnt bleiben, daß Anhaltspunkte für weitere Funktionen des Carnitins bestehen. So erscheint eine Mitwirkung bei der Corticosteroidsynthese in den Nebennieren und eine Teilnahme bei Vorgängen der Nervenerregung möglich [146]. Interessante Ergebnisse sind auch von der weiteren Erforschung der Rolle des Carnitins bei der männlichen Fertilität zu erwarten. In ähnlicher Weise, wie der Fructosegehalt im Sperma von der Produktion der androgenen Hormone abhängt, wird auch der Gehalt an Carnitin und Carnitin-Acetyltransferase von diesen Hormonen beeinflußt [91]. Eine Prüfung, ob dieses Verhalten ähnlich wie bei der Fructose unsere diagnostischen Möglichkeiten bei Fertilitätsstörungen erweitern kann, erscheint lohnend.

Zusammenfassung

Es werden zuerst das Vorkommen des Carnitins in verschiedenen Organen seine Biosynthese und Turnover besprochen. Am Zellstoffwechsel der Fettsäuren kann das Carnitin teilnehmen, da es durch eine kurzkettige und eine langkettige Transferase acyliert werden kann. Die Acylcarnitine stellen neben den schon lange bekannten Acyl-CoA-Thioestern eine weitere Form aktivierter, also energiereicher Fettsäuren dar. Die Acyle beider Stoffklassen sind in der Zelle über die Transferasen austauschbar und wahrscheinlich miteinander im Gleichgewicht. Daraus folgt, daß z. B. der Pool der aktivierten Essigsäure das Acetyl-CoA und das Acetylcarnitin umfaßt. Die Bedeutung des Carnitins liegt in seiner Fähigkeit (im Gegensatz zu den Acyl-Thioestern), Zellmembranen gut penetrieren zu können und so als Carrier für aktivierte kurz- und langkettige Fettsäuren wirken zu können. Dabei scheint die Carrierfunktion nicht nur auf den Austausch zwischen extra- und intramitochondrialen Kompartimenten bedeutsam zu sein, sondern auch zwischen verschiedenen intramitochondrialen Räumen zu wirken.

Nach Besprechung der Pathophysiologie der aktivierten Acyle und des Quotienten Acyl-CoA/freies CoA wird die Rolle des Carnitins in diesem System besprochen. Bei vermehrtem Angebot an Fettsäuren mit entsprechend verschobenem Quotienten Acyl-CoA/freies CoA, wie es z.B. bei der diabetischen Stoffwechselentgleisung vorkommt, ergibt sich die Möglichkeit, durch Carnitinzufuhr therapeutisch einzuwirken. Carnitin stellt pharmakologisch eine sehr gering toxische, körpereigene Verbindung dar. Es ist der einzige bisher bekannte Stoff, mit dem eine Veränderung des Grades der Acylierung von Coenzym A erreicht werden kann. Die theoretischen Grundlagen für einen Einsatz des Carnitins bei der diabetischen Acidose werden erörtert. Abschließend wird auf das androgenabhängige Verhalten des Carnitins im Sperma bzw. in den männlichen Geschlechtsdrüsen hingewiesen, die sich diagnostisch bei Fertilitätsstörungen verwerten lassen könnten.

Literatur

1. ABDEL-KADER, M. M., WOLF, G.: The distribution of carnitine and its possible function in corticosteroid biosynthesis. In: WOLF, G. (ed.) [146], p. 147—156 (1964).
2. ALLMANN, D. W., GALZIGNA, L., McCAMAN, R. E., GREEN, D. E.: The membrane systems of the mitochondrion. IV. The localization of the fatty acid oxidizing system. Arch. Biochem. 117, 413—422 (1966).
3. — HARRIS, R. A., GREEN, D. E.: Site of action of atractyloside in mitochondria. I. Inhibition of outer membrane functions by atractyloside. Arch. Biochem. 120, 693—702 (1967).
4. BALLARD, F. B., DANFORTH, W. H., NAEGLE, S., BING, R. J.: Myocardial metabolism of fatty acids. J. clin. Invest. 39, 717—729 (1960).
5. BEENAKERS, A. M. TH., KLINGENBERG, M.: Carnitine coenzyme A transacetylase in mitochondria from various organs. Biochim. biophys. Acta (Amst.) 84, 205—207 (1964).
6. BENDER, A. E., ADAMS, E. P.: An investigation of suggested physiological functions of carnitine. Biochem. J. 82, 232—236 (1962).
7. BENMILOUD, M., FREINKEL, N.: Stimulation of gluconeogenesis by carnitine in vitro. Metabolism 16, 658—669 (1967).
8. BICKEL, A., KORCHOV, A.: Beitrag zur Kenntnis des chemischen Charakters der Sekretinstoffe in den Nahrungsmitteln. Biochem. Z. 199, 434—444 (1928).
9. BLIXENKRONE-MØLLER, N.: Respiratorischer Stoffwechsel und Ketonbildung der Leber. Hoppe-Seylers Z. physiol. Chem. 252, 117—136 (1938).
10. BODE, CH., KLINGENBERG, M.: Die Veratmung von Fettsäuren in isolierten Mitochondrien. Biochem. Z. 341, 271—299 (1965).
11. BØHMER, T., NORUM, K. R., BREMER, J.: The relative amounts of long-chain acylcarnitine, acetylcarnitine, and free carnitine in organs of rats in different nutritional states and with alloxan diabetes. Biochim. biophys. Acta (Amst.) 125, 244—251 (1966).
12. — Tissue levels of activated fatty acids (acylcarnitines) and the regulation of fatty acid metabolism. Biochim. biophys. Acta (Amst.) 144, 259—270 (1967).
13. BORTZ, W. M., LYNEN, F.: Elevation of lang chain acyl-CoA derivatives in livers of fasted rats. Biochem. Z. 339, 77—82 (1963).
14. — — The inhibition of acetyl CoA carboxylase by long chain acyl CoA derivatives. Biochem. Z. 337, 505—509 (1963).
15. BREMER, J.: Biosynthesis of carnitine in vivo. Biochim. biophys. Acta (Amst.) 48, 622—624 (1961).
16. — Carnitine precursors in the rat. Biochim. biophys. Acta (Amst.) 57, 327—335 (1962).

17. BREMER, J.: Carnitine in intermediary metabolism. Reversible acetylation of carnitine by mitochondria. J. biol. Chem. 237, 2228—2231 (1962).

18. — Carnitine in intermediary metabolism. The metabolism of fatty acid esters of carnitine by mitochondria. J. biol. Chem. 237, 3628—3632 (1962).

19. — Carnitine in intermediary metabolism. The biosynthesis of palmitylcarnitine by cell subfractions. J. biol. Chem. 238, 2774—2779 (1963).

20. — Comparison of acylcarnitines and pyruvate as substrates for rat-liver mitochondria. Biochim. biophys. Acta (Amst.) 116, 1—11 (1966).

21. — NORUM, K. R.: Palmityl-CoA: carnitine O-palmityltransferase in the mitochondrial oxidation of palmityl-CoA. Europ. J. Biochem. 1, 427—433 (1967).

22. BRESSLER, R., KATZ, R. J.: The role of carnitine in acetoacetate production and fatty acid synthesis. In: WOLF, G. (146), p. 65—81 (1964).

23. — — The role of carnitine in acetoacetate production and fatty acid synthesis. J. clin. Invest. 44, 840—849 (1965).

24. BRESSLER, R., KATZ, R. J.: The effect of carnitine on the rate of incorporation of precursors into fatty acids. J. biol. Chem. 240, 622—627 (1965).

25. — BRENDEL, K.: The role of carnitine and carnitine acyltransferase in biological acetylations and fatty acid synthesis. J. biol. Chem. 241, 4092—4097 (1966).

26. BROEKHUYSEN, J., DELTOUR, G.: Teneur en carnitine de quelques milieux biologiques. Ann. Biol. clin. 19, 549—558 (1961).

27. — BAUDINE, A., DELTOUR, G.: Effect of carnitine on acidosis and ketosis induced by lipid perfusions in dog during starvation. Biochim. biophys. Acta (Amst.) 106, 207—210 (1965).

28. — ROZENBLUM, C., GHISLAIN, M., DELTOUR, G.: Distribution of carnitine in the rat. In: WOLF, G. (146), p. 23—25 (1964).

29. CHAPPELL, J. B., CROFTS, A. R.: The effect of atractylate and oligomycin on the behaviour of mitochondria towards adenine nucleotides. Biochem. J. 95, 707—716 (1965).

30. CHASE, J. F. A., PEARSON, D. J., TUBBS, P. K.: The preparation of crystalline carnitine acetyltransferase. Biochim. biophys. Acta (Amst.) 96, 162—165 (1965).

31. — pH-dependence of carnitine acetyltransferase activity. Biochem. J. 104, 503—509 (1967).

32. — The substrate specificity of carnitine acetyltransferase. Biochem. J. 104, 510—518 (1967).

33. CHILDRESS, CH. C., SACKTOR, B., TRAYNOR, D. R.: Function of carnitine in the fatty acid oxidase deficient flight muscle. J. biol. Chem. 242, 754—760 (1967).

34. CORREDOR, C., MANSBACH, C., BRESSLER, R.: Carnitine depletion in the choline-deficient state. Biochim. biophys. Acta (Amst.) 144, 366—374 (1967).

35. CRUICKSHANK, E. W. H., STARTUP, C. W.: The effect of insulin on the respiratory quotient, oxygen consumption, sugar utilization, and glycogen synthesis in the normal mammalian heart in hyper- and hypoglycaemia. J. Physiol. (Lond.) 77, 365—385 (1933).

36. ERFLE, J. D., SAUER, F.: Acetyl coenzyme A and acetylcarnitine concentration and turnover rates in muscle and liver of the ketotic rat and guinea pig. J. biol. Chem. 242, 1988—1996 (1967).

37. EVANS, J. R., OPIE, L. H., RENOLD, A. E.: Pyruvate metabolism in the perfused rat heart. Amer. J. Physiol. 205, 971—976 (1963).

38. FRAENKEL, G.: Studies on the distribution of vitamin B_T (Carnitine). Biol. Bull. 104, 359—371 (1953).

39. — FRIEDMAN, S., HINTON, T., LASZLO, S., NOLAND, J.: The effect of substituting carnitine for choline in the nutrition of several organisms. Arch. Biochem. 54, 432—439 (1955).

40. — — Carnitine. Vitam. and Horm. 15, 73—118 (1957).

41. Friedman, A. D., Rumsay, P., Graff, S.: The metabolism of pyruvate in the tricarboxylic acid cycle. II. Tissue characteristic metabolism of pyruvate. J. biol. Chem. **235**, 1854—1855 (1960).
42. Friedberg, S. J., Bressler, R.: The formation and isolation of longchain acyl-carnitines in mitochondria. Biochim. biophys. Acta (Amst.) **98**, 335—343 (1965).
43. Friedman, S., Fraenkel, G.: Reversible enzymatic acetylation of carnitine. Arch. Biochem. **59**, 491—501 (1955).
44. — Determination of carnitine in biological materials. Arch. Biochem. **75**, 24—30 (1958).
45. Fritz, I. B.: The effects of muscle extracts on the oxidation of palmitic acid by liver slices and homogenates. Acta physiol. scand. **34**, 367—385 (1955).
46. — Effects of choline deficiency and carnitine on palmitic acid oxidation by rat liver homogenates. Amer. J. Physiol. **190**, 452—499 (1957).
47. — Davis, D. G., Holtrop, R. H., Dundee, H.: Fatty acid oxidation by skeletal muscle during rest and activity. Amer. J. Physiol. **194**, 379—386 (1958).
48. — Action of carnitine on long chain fatty-acid oxidation by liver. Amer. J. Physiol. **197**, 297—304 (1959).
49. — Kaplan, E., Yue, K. T. N.: Specificity of carnitine action on fatty acid oxidation by heart muscle. Amer. J. Physiol. **202**, 117—121 (1962).
50. — Yue, K. T. N.: The site of carnitine action on long-chain fatty acid oxidation: a possible new pathway for acyl CoA formation in heart muscle. Physiologist **5**, 144—149 (1962).
51. — Schultz, S. K., Srere, P. A.: Properties of partially purified carnitine acetyl-transferase. J. biol. Chem. **238**, 2509—2517 (1963).
52. — Yue, K. T. N.: Long-chain carnitine acyltransferase and the role of acylcarnitine derivatives in the catalytic increase of fatty acid oxidation induced by carnitine. J. Lipid Res. **4**, 279—288 (1963).
53. — Possible role of carnitine acetyltransferase in mitochondrial translocation of acetyl groups during fatty acid synthesis from glucose by adipose tissue. In: Wolf, G. (ed.) [146], p. 83—86 (1964).
54. — Closing remarks. In: Wolf, G. (ed.) [146], p. 199—204 (1964).
54a. — Yue, K. T. N.: Effects of carnitine on acetyl-CoA oxidation by heart muscle mitochondria. Amer. J. Physiol. **206**, 531—535 (1964).
55. — Schultz, S. K.: Carnitine acetyltransferase. II. Inhibition by carnitine analogues and by sulfhydryl reagents. J. biol. Chem. **240**, 2188—2193 (1965).
56. — Relief of palmityl CoA inhibition of citrate synthase by long chain acylcarnitine derivatives. Biochim. biophys. Res. Commun. **22**, 744—748 (1966).
57. Garkavi, P. G.: Über die Rolle des Carnitins in einigen biochemischen Prozessen und sein Gehalt im Gewebe einiger Tiere [Russ.]. Biokhimiya **18**, 302—304 (1953).
58. Garland, P. B., Shepherd, D., Yates, D. W.: Steady-state concentrations of coenzyme A, acetyl-coenzyme A and long-chain fatty acyl coenzyme A in rat-liver mitochondria oxidizing palmitate. Biochem. J. **97**, 587—594 (1965).
59. Geyer, R. P., Mathews, L. W., Starl, F. J.: Metabolism of emulsified trilaurin (—C¹⁴OO—) and octanoic acid (—C¹⁴OO—) by rat tissue slices. J. biol. Chem. **180**, 1037—1045 (1949).
60. Gordon, R. S., Jr.: Unesterified fatty acid in human blood plasma. II. The transport function of unesterified fatty acid. J. clin. Invest. **36**, 810—815 (1957).
61. Greville, G. D.: Factors affecting the utilization of substrates by mitochondria. In: Tager et al. (126), p. 86—106 (1966).
62. Gulewitsch, V. S., Krimberg, R.: Zur Kenntnis der Extraktivstoffe der Muskeln. II. Mitt. Über das Carnitin. Hoppe-Seylers Z. physiol. Chem. **45**, 326—330 (1905).
63. Haft, D. E.: Evidence for inhibition of acetyl-coenzyme A formation pyruvate in diabetic rat liver. Biochim. biophys. Acta (Amst.) **90**, 173—175 (1964).

64. HALL, L. M.: Preferential oxidation of acetoacetate by perfused heart. Biochim. biophys. Res. Commun. **6**, 177—179 (1961).

65. HOLZER, H.: Diskussionsbemerkung zu THOMITZEK, W.-D. (129). Verh. Ges. exp. Med. DDR, Bd. 6 (Tagg Jena 17.—19. 10. 1963), S. 362. Dresden u. Leipzig: Th. Steinkopff 1964.

66. HÜLSMANN, W. C., SILIPRANDI, D., CIMAN, M., SILIPRANDI, N.: Effect of carnitine on the oxidation of α-oxoglutarate to succinate in the presence of acetoacetate or pyruvate. Biochim. biophys. Acta (Amst.) **93**, 166—168 (1964).

67. — WIT-PEETERS, E. M., BENKHUYSEN, C.: Factors influencing fatty acid metabolism in mitochondria. In: TAGER, J. M. (126), p. 460—474 (1966).

68. JAENICKE, L., LYNEN, F.: Coenzyme A. In: The enzymes (P. D. BOYER, H. LARDY and K. MYRBÄCK, ed.), vol. III, p. 3—105. New York: Academic Press. Inc. 1960.

69. KEECH, D. B., UTTER, M. F.: Pyruvate carboxylase. II. Properties. J. biol. Chem. **238**, 2609—2614 (1963).

70. KENNEDY, E., LEHNINGER, A. L.: Oxidation of fatty acids and tricarboxylic acid cycle intermediates by isolated rat liver mitochondria. J. biol. Chem. **179**, 957—965 (1949).

71. — The products of oxidation of fatty acids by isolated rat liver mitochondria. J. biol. Chem. **185**, 275—285 (1950).

72. KHAIRALLAH, E. A., MEHLMAN, M. M.: The turnover, body pool, and daily excretion of carnitine as determined by isotopedilution technique. In: WOLF, G. (ed.) [146], p. 57—62 (1964).

73. — WOLF, G.: Carnitine decarboxylase. J. biol. Chem. **242**, 32—39 (1967).

74. KORCHOV, A.: Über die Wirkung einiger nach dem Verfahren von GULEWITSCH und KRIMBERG gewonnener Fraktionen des Liebigschen Fleischextraktes auf die Magensekretion. Biochem. Z. **190**, 188—198 (1927).

75. KROGH, A., WEISS-FOGH, T.: The respiratory exchange of the desert locust (Schistocerca gregaria) before, during and after flight. J. exp. Biol. **28**, 344—357 (1951).

76. KUNZE, D., STRACK, E.: Über die Wirkung von Carnitin und Acetylcarnitin auf die Acetacetatbildung in Mäuseleber. Acta biol. med. germ. **13**, 832—841 (1964).

77. KUTSCHER, F.: Über Liebig's Fleischextrakt. Z. Untersuch. Nähr-. u. Genußmitt. **10**, 528—537 (1905).

78. LELOIR, L. F., MUNOZ, J. M.: Fatty acid oxidation in liver. Biochem. J. **33**, 734—746 (1939).

79. LINDSTEDT, G., LINDSTEDT, S.: On the biosynthesis and degradation of carnitine. Biochem. biophys. Res. Commun. **6**, 319—323 (1961).

80. LINDSTEDT, S., LINDSTEDT, G.: Distribution and excretion of carnitine in the rat. Acta chem. scand. **15**, 701—702 (1961).

81. LINDSTEDT, G., LINDSTEDT, S.: On the hydroxylation of γ-butyrobetaine to carnitine in vitro. Biochem. biophys. Res. Commun. **7**, 394—397 (1962).

82. — — Studies on the biosynthesis of carnitine. Biochem. J. **84**, 84 P (1962).

83. — — Studies on the biosynthesis of carnitine. J. biol. Chem. **240**, 316—321 (1965).

84. — Effect of metal ions on the hydroxylation of γ-butyrobetaine to carnitine in rat-liver homogenates. Biochim. biophys. Acta (Amst.) **141**, 492—498 (1967).

85. LINNEWEH, W.: γ-Butyrobetain, Crotonbetain und Carnitin im tierischen Stoffwechsel. Hoppe-Seylers Z. physiol. Chem. **181**, 42—53 (1929).

86. LOSSOW, W. J., BROWN, G. W., CHAIKOFF, I. L.: Sparing of palmitic acid oxidation by carbohydrate: Prefeeding versus addition to medium. J. biol. Chem. **222**, 531—535 (1956).

87. LOWENSTEIN, J. M.: A comparison of acetyl group utilization from various precursors. In: WOLF, G. (146), p. 97—112 (1964).

88. LYNEN, F.: Lipid metabolism. Ann. Rev. Biochem. **24**, 653—688 (1955).

89. MARQUIS, N. R., FRITZ, I. B.: Enzymological determination of free carnitine concentrations in rat tissues. J. Lipid Res. **5**, 184—187 (1964).

90. Marquius, N. R., Fritz, I. B.: The distribution of carnitine, acetylcarnitine, and carnitine acetyltransferase in rat tissues. J. biol. Chem. **240**, 2193—2196 (1965).

91. — — Effects of testosterone on the distribution of carnitine, acetylcarnitine, and carnitine acetyltransferase in tissues of the reproductive system of the male rat. J. biol. Chem. **240**, 2197—2200 (1965).

92. McCaman, R. R., McCaman, M. W., Stafford, M. L.: Carnitine acetyltransferase in nervous tissue. J. biol. Chem. **241**, 930—934 (1966).

93. Mehlman, M. A., Wolf, G.: Studies on the distribution of free carnitine and the occurence and nature of bound carnitine. Arch. Biochem. **98**, 146—153 (1962).

94. — — Phosphatidylcarnitine. Arch. Biochem. **102**, 346—354 (1963).

95. Meyer, H., Preiss, B., Bauer, S.: The oxidation of fatty acids by a particulate fraction from desert-locust (Schistocerca gregaria) thorax tissues. Biochem. J. **76**, 27—35 (1960).

96. Monod, J., Changeux, J.-P., Jacob, F.: Allosteric proteins and cellular control systems. J. molec. Biol. **6**, 306—329 (1963).

97. Montgomery, C. M., Webb, J. L.: Metabolic studies on heart mitochondria. I. The operation of the normal tricarboxylic acid cycle in the oxidation of pyruvate. J. biol. Chem. **221**, 347—357 (1956).

98. Norum, K. R.: Palmityl-CoA: carnitine palmityltransferase. Purification from calf-liver mitochondria and some properties of the enzyme. Biochim. biophys. Acta (Amst.) **89**, 95—108 (1964).

99. — Activation of palmityl-CoA: carnitine palmityl-transferase in livers from fasted, fat-fed, or diabetic rats. Biochim. biophys. Acta (Amst.) **98**, 652—654 (1965).

100. — Bremer, J.: The localization of acyl coenzyme A-carnitine acyltransferases in rat liver cells. J. biol. Chem. **242**, 407—411 (1967).

101. Olson, R. E.: Effect of pyruvate and acetoacetate on the metabolism of fatty acids by the perfused rat heart. Nature (Lond.) **195**, 597—599 (1962).

102. Pearson, D. J., Tubbs, P. K.: Acetyl-carnitine in heart and liver. Nature (Lond.) **202**, 91 (1964).

103. — — Tissue levels of acid-insoluble carnitine in rat heart. Biochim. biophys. Acta (Amst.) **84**, 772—773 (1964).

104. Plaut, G. W., Plaut, K. A.: Oxidative metabolism of heart mitochondria. J. biol. Chem. **199**, 141—151 (1952).

105. Purvis, J. L., Lowenstein, J. M.: The relation between intra- and extramitochrondrial pyridine nucleotides. J. biol. Chem. **236**, 2794—2803 (1961).

106. Randle, P. J., Garland, P. B., Hales, C. N., Newsholme, E. A.: The glucose fatty acid cycle and diabetes mellitus. Ciba found. Coll. Endocr. **15**, 192—212 (1964).

107. Reynier, M.: Contribution à l'étude théorique et expérimentale des propriétés biologiques de la carnitine. Rev. Agressol. **4**, 361—373 (1963).

108. — Action antagoniste de la carnitine vis-à-vis de l'acidose du jeûne chez de lapin. Rev. Agressol. **4**, 505—507 (1963).

109. Rossi, C. R., Galzigna, L., Gibson, D. M.: Fatty acid activation and oxidation in rat liver mitochondria. In: Tager, J. M., et al. (126), p. 143—152 (1966).

110. Rotzsch, W., Lorenz, I., Strack, E.: Über die Toxizität des Carnitins und einiger verwandter Stoffe. Acta biol. med. germ. **3**, 28—36 (1959).

111. Sauer, F., Erfle, J. D.: On the mechanism of acetoacetate synthesis by guinea pig liver fractions. J. biol. Chem. **241**, 30—37 (1966).

112. Severin, S. E.: Diskussionsbemerkung zu Thomitzek, W.-D. (129). Verh. Ges. exp. Med. DDR, Bd. 6 (Tagg Jena 17.—19. 10. 1963). Dresden u. Leipzig: Th. Steinkopff 1964, S. 362.

113. Shepherd, D., Yates, D. W., Garland, P. B.: The relationship between the rates of conversion of palmitate into citrate or acetoacetate and the acetyl-coenzyme A content of rat-liver mitochondria. Biochem. J. **97**, 38 C—40 C (1965).

114. SODERBERG, J., THERRIAULT, D. G., WOLF, G.: Characterization of lipid-bound carnitine. In: WOLF, G. (146), p. 165—171 (1964).

115. SPENCER, A. F., LOWENSTEIN, J. M.: The supply of precursors for the synthesis of fatty acids. J. biol. Chem. 237, 3640—3648 (1962).

116. STRACK, E., FÖRSTERLING, K.: Über die biologische Wirkung des Carnitins und Azetylcarnitins. Naunyn-Schmiedebergs Arch. exp. Path. Pharmak. 185, 612—621 (1937).

117. — — Über die Reizkraft der Methylester einiger Betaine. Hoppe-Seylers Z. physiol. Chem. 285, 207—216 (1950).

118. — — Über die Reizwirkungen von Estern des l-Carnitins auf isolierte Organe. Hoppe-Seylers Z. physiol. Chem. 295, 377—387 (1953).

119. — LORENZ, I.: Zur Bestimmung des Carnitins. Hoppe-Seylers Z. physiol. Chem. 298, 27—33 (1954).

120. — — Die Darstellung von L-Carnitin und seiner Isomeren. Hoppe-Seylers Z. physiol. Chem. 318, 129—137 (1960).

121. — — Zur Darstellung von O-Acyl-Derivaten des Carnitins. Hoppe-Seylers Z. physiol. Chem. 343, 231—239 (1966).

122. — KUNZE, D., MÜLLER, H.-P.: Über den Einfluß von Carnitin auf die Keton-körperbildung in vitro und in vivo. Acta biol. med. germ. 12, 716—718 (1964).

123. — Klinische Anwendung von Carnitin bei Hyperthyreose. Dtsch. Gesund.-Wes. 22, 2055—2059 (1967).

124. STRENGTH, D. R., YU, S. Y.: Origin of methyl group of carnitine. Fed. Proc. 21, 1 (1962).

125. — — DAVIS, E. Y.: Biosynthesis of carnitine, dietary factors that influence concen-tration of carnitine in tissues. In: WOLF, G. (ed.) [146], p. 45—56 (1964).

126. TAGER, J. M., PAPA, S., QUAGLIARIELLO, E., SLATER, E. C. (eds): Regulation of metabolic processes in mitochondria. Biochim. biophys. Acta Library, vol. 7. Amsterdam-London-New York: Elsevier Publ. Co. 1966.

127. TAKETA, K., POGELL, B. M.: The effect of palmityl coenzyme A on glucose 6-phos-phate dehydrogenase and other enzymes. J. biol. Chem. 241, 720—726 (1966).

128. THERRIAULT, D. G., MEHLMAN, M. A.: Carnitine-C¹⁴ metabolism in cold acclimated rats. In: WOLF, G. (ed.) [146], p. 141—145 (1964).

129. THOMITZEK, W.-D.: Die Wirkung von Derivaten des (—)- und (+)-Carnitins und Acetylcarnitins auf die Bildung von Acetylcholin. Biochem. Pharmacol., Suppl. 12, Conference Issue, 2nd Int. Pharmacol. Meeting, Prague 1963, p. 161.

130. — Der Einfluß von Derivaten der γ-Amino-β-hydroxybuttersäure auf fermentative Transazetylierungen. Verh. Ges. exp. Med. DDR, Bd. 6 (Tagg Jena 17.—19. 10. 1963). Dresden u. Leipzig: Th. Steinkopff 1964, S. 358—361.

131. — STRACK, E.: Acetylcarnitin als Acetyldonator bei der enzymatischen Bildung von Acetylcholin. Acta biol. med. germ. 13, 110—125 (1964).

132. — WILLGERODT, H.: Über die Penetration von verschiedenen quaternären N-Ver-bindungen in Erythrocyten und ihren Einfluß auf die Erythrocytenmembran. Folia haemat. (Lpz.) 78, 58—63 (1965).

133. — WINTER, H., STRACK, E.: Über den Einfluß von Carnitin auf den Azetylcholin-gehalt in Gehirn und Herzmuskel in vivo und auf die Azetylierung von Sulfanilamid in der Leber. Acta biol. med. germ. 16, 350—358 (1966).

134. TOMITA, M., SENDJU, Y.: Über die Oxyaminoverbindungen, welche die Biuretreak-tion zeigen. III. Spaltung der γ-Amino-β-oxybuttersäure, in die optisch-aktiven Komponenten. Hoppe-Seylers Z. physiol. Chem. 169, 263—277 (1927).

135. TUBBS, P. K.: Inhibition of citrate formation by long chain acyl thio esters of coenzyme A as possible control mechanism in fatty acid biosynthesis. Biochim. biophys. Acta (Amst.) 70, 608—609 (1963).

136. — GARLAND, P. B.: Fatty acyl thio esters of coenzyme A: Inhibition of fatty acid synthesis in vitro and determination of levels in liver in normal. fasted, and fat- or sugar-fed rats. Biochem. J. 89, 25 P (1963).

137. TUBBS, P. K., GARLAND, P. B.: Variations in tissue contents of coenzyme A thio esters and possible metabolic implications. Biochem. J. **93**, 550—557 (1964).
138. UTTER, M. F., KEECH, D. B.: Pyruvate carboxylase. I. Nature of reaction. J. biol. Chem. **238**, 2603—2608 (1963).
139. VAN DEN BERGH, S. G.: Diskussionsbemerkung in: TAGER, J. M., et al. (eds.) [126], p. 106 (1966).
140. VERLY, W. G., BACQ, Z. M.: Prise oralement la «dicarnitine» racémique n'est pas un bon donneur de méthyle pour le rat. Biochim. biophys. Acta (Amst.) **13**, 454—455 (1954).
141. VOLK, M. E., MILLINGTON, R. H., WEINHOUSE, S.: Oxidation of endogenous fatty acids of rat tissues in vitro. J. biol. Chem. **195**, 493—501 (1952).
142. WIELAND, O., WEISS, L.: Inhibition of citrate-synthase by palmityl-coenzyme A. Biochem. biophys. Res. Commun. **13**, 26—31 (1963).
143. — — EGER-NEUFELDT, I., TEINZER, A., WESTERMANN, B.: Coenzym A-Thioester höherer Fettsäuren als mögliche Vermittler enzymatischer Regulationen im Tierkörper. Klin. Wschr. **43**, 645—654 (1965).
144. WILLIAMSON, J. R., KREBS, H. A.: Acetoacetate as fuel of respiration in the perfused rat heart. Biochem. J. **80**, 540—547 (1961).
145. WOLF, G., BERGER, C. R. A.: Studies on the biosynthesis and turnover of carnitine. Arch. Biochem. **92**, 360—365 (1961).
146. — (ed.): Recent research on carnitine. Its relation to lipid metabolism (Symposium 24./25. 7. 1964). Cambridge (Mass.): M. I. T. Press 1964.
147. YUE, K. T. N., FRITZ, I. B.: Fate of tritium-labeled carnitine administered to dogs and rats. Amer. J. Physiol. **202**, 122—128 (1962).

Addendum

Während der Drucklegung des Manuskripts sind in der Literatur Ergebnisse mitgeteilt worden, die die vorstehenden Ausführungen ergänzen.

WILLIAMSON et al. [Diabetes **17**, 194—208 (1968)] konnten mit dem Hemmstoff der (—)-Carnitin-Palmityltransferase (+)-Decanoylcarnitin an der perfundierten Rattenleber zeigen, daß in der Leber die langkettigen Fettsäuren überwiegend durch carnitinabhängige Wege oxydiert werden, während die kürzerkettigen (z. B. Octanoat) nicht carnitinabhängig verwertet werden.

BARKER et al. [Biochem. J. **110**, 739—746 (1968)] fanden, daß die bekanntlich mitochondrial lokalisierte Carnitin-Acetyltransferase in vivo für extramitochondriales Acetyl-CoA nicht oder nur zu einem sehr kleinen Prozentsatz zur Verfügung steht. Frühere Versuche [24, 54a] hatten Carnitin-Acetyltransferase eine wichtige Rolle als Mittler zwischen extra- und intramitochondrialem CoA zugeschrieben. Die quantitative Abschätzung zeigt aber, daß bereits etwa 1% der Gesamtaktivität ausreicht, um diese Ergebnisse [24, 54a] zu erklären. Damit stellt sich die Frage nach der Hauptrolle der Carnitin-Acetyltransferase erneut in vollem Umfang. GRAVINA und GRAVINA-SANVITALE [Clin. chim. Acta **23**, 376—377 (1969)] haben die antiketotische Wirkung von Carnitin bestätigt. Der erhöhte Acetoacetatspiegel im Blut fastender Kinder fiel nach intravenöser Carnitingabe ab. Dieser Abfall könnte durch Senkung der Ketokörperbildung und nicht durch Steigerung der peripheren Verwertung erfolgen. Dann läßt sich auch leicht erklären, daß SÖLING und APPELS [Biochim. biophys. Acta (Amst.) **158**, 162—164 (1968)], die die periphere Utilisation von zugeführtem Acetoacetat an eviscerierten Ratten gemessen haben, eine gegenteilige Carnitinwirkung fanden. Carnitin scheint also eine antiketotische Wirkung nur bei erhöhter Ketokörperbildung zu entfalten.

Today's Carcinochemotherapy: Some of its Achievements, Failures and Prospects*

Franz Bergel

With 10 Figures

Table of Contents

I. Introduction

Morbid definition of cancer. If someone talks or writes about a therapy applied to a specific disease, he assumes that the disease is well defined from a pathological and etiological point of view and that the mechanisms of the causation, and reaction to treatments, if not fully elucidated, are at least known to a considerable extent. When he confronts "cancer" with the same set of enquiries, he discovers quickly that it is not *one* disease. Though the fact that "it" is a cellular disease joins a great number of the neoplastic

---

* This short review is dedicated to Sidney Farber and his Children's Cancer Research Foundation where, in 1947, carcinochemotherapy was initiated.

groups and sub-groups together, one of these neoplasms is as different from the other as measles from influenza, in spite of the accepted view that both these illnesses are caused by viruses. There is another reason, namely the multi-etiology of many cancers, which makes it very difficult to pronounce at the very beginning of this article that, while there is obviously, at this moment, no one cancer cure, it may be round the next corner. In consequence these two, often ignored properties, morbid pluralism and multi-etiology, hinder attempts to produce a relatively simple story of therapy, when treatment of a nonsurgical and non-radiation type is the main theme. It is the author's opinion that pluralistic pathology and multi-etiology make any such task like that of a politician who promises many things before he is in power and can only fulfill very few when he is in the saddle. Moreover, when attempting to review such difficult fields, it is thought, more often than not, that it is the wrong moment to prepare a critical survey. "Let us wait" is just as wrong a counsel as the demand for a solid, encyclopaedic basis on which to rest comments, criticism and prognosis.

Available treatments. In order to clarify at the very start all kinds of treatments available to the medical profession in the case of cancerous diseases (see MANUILA et al., 1967; MATHÉ, 1969a; BOESEN and DAVIS, 1969) table 1 is presented; it indicates existing and potential modes of therapies (see BERGEL, 1967).

An important division between therapies is determined by the mode in which neoplastic diseases develop. If they remain anatomically localised, then surgery and radiotherapy, both of which have improved in their technical accuracy over the years (see MANUILA, 1967), can deal with these neoplasms to a large extent. But if the diseases are disseminated over large areas like most forms of leukemias or through metastases of tumours, then types of therapy have to be chosen which act either systemically or which are capable of exerting widespread effects. They are available in the form of anti-cancer drugs and, possibly more so in the future, of a number of chemo-biological variants. They are represented by immunological, anti-viral or restitutional therapy, the last embracing methods which aim at re-forming equilibria that have been disturbed. Naturally, the most important efforts should be directed toward improving the selectivity of all these remedies; this means that they should act as specifically as pharmacodynamic agents but within the narrow confines of the cancerous disease to be treated.

Prevention. We shall come back to the present status of each of these modes of therapy; however, before doing so, it is necessary and logical to investigate "prevention". This means in the case of neoplastic diseases (in contrast to infectious diseases which force the individual to stay away from certain areas where the illness is endemic or epidemic), the avoidance of contact with a great number of carcinogenic agents and materials. Apart from those of a purely physical nature, i.e. injury, irritation, cold, irradiation, such as U.V. and X-ray, they consist of chemicals, viruses (their effect on human beings, strangely enough, is not established yet beyond doubt) and biochemical lesions, most of which may possibly be connected with congenital mishaps. Because of the internal or endogenous nature of the latter, it is hardly likely that one can escape them. If either as an individual or as a

Table 1

Prevention

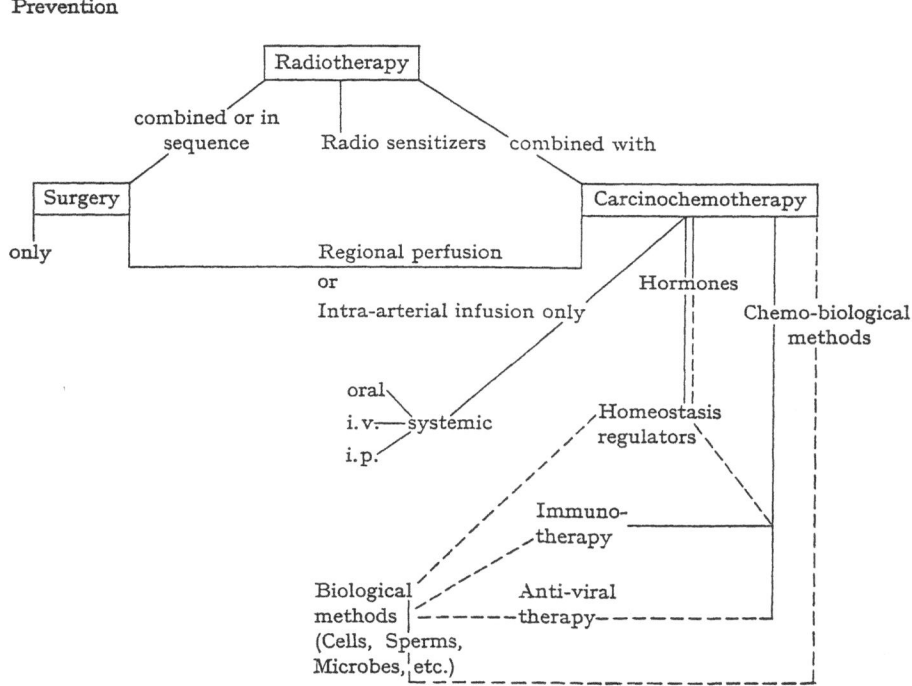

member of a community, governed by reasonable regulations of hygiene, nutrition and environmental contaminations, one avoids or diminishes contacts with sources of carcinogenic agents or with the agents themselves, then these preventive measures, if systematically applied, may be given the name of "negative chemotherapy", in analogy to BOYLAND's (1961) name for hormone withdrawal.

Multiple etiology — multiple treatment. It looks, at present, as if the causation of nearly all the cellular aberrations, called neoplastic or malignant diseases, are of a multiple character, either with respect to the causative agents or to the cellular receptive sites which, after impact of the agents or the lesions inflicted by them, may respond with aberrant reactions, leading to cancer. At the cellular level these sites can be the nucleus, or the subcellular components such as mitochondria, lysosomes, surface and internal membranes, a discrete matter of the cytoplasm, or all together (BERGEL, 1966a).

It is not yet established with certainty whether the end effects of such sequentially or simultaneously occurring perturbations or injuries differ from organ to organ and from organism to organism or individual to individual. It is not improbable that only the mechanisms differ, while the final changes, in a biological and biochemical sense, may be identical. Such changes by establishing themselves permanently through template replication could in time lead to the development of malignancies. This, so it is agreed, might happen on a single cell basis, in spite of the objections SMITHERS (1964) raised in his book "On the Nature of Neoplasia in Man" which favours an involvement of the whole organism. If the latter hypothesis were correct, then for that reason alone a multiple approach by chemotherapy would be hardly justifiable. Taking into account the small number of known effective cancer drugs, this would require the application of different mixtures, different routes of administrations, and a long list of combinations with other forms of treatments.

If, on the other hand, the numerous stages of carcinogenesis show subtle differences even between individual patients, or if, apart from the multi-etiology of the cancers, the slowly unravelled mechanisms point to differences in the sites of primary lesions, then an intensive search for a more rational and controllable approach becomes a matter of urgency. But then it will be of importance whether neoplastic diseases are closely related to parasitical states or not, allowing for the endogenous nature of these "parasites". The reason for this is that, in case the "parasitical" behaviour dominates, any inherent or acquired resistance against treatment may follow rules similar to those governing microbial diseases. Because parasitic organisms and their cells can never be amalgamated with or transformed into host tissues and cells, although perhaps a kind of symbiosis might develop, the only way to get rid of them is by complete destruction.

There exists a respectable number of believers in the uni-etiological origin and irreversibility of cancerous states. As promulgated by WARBURG (1966) on several occasions, he is convinced, on the basis of his own observations, that these two premises are correct. Well, the monoetiology will not prevent successful chemotherapy if the therapist uses the total eradication approach. Unfortunately, however, the idea that cancer is completely irreversible is scientifically and intellectually holding off any attempt at restitutional work, that is the reversal of all the processes that have led to the neoplastic and de-differentiated state of cells and tissues. Whether such a reversal back to the mature cell, in complete possession of its regulating powers, is experimentally and then practically possible, has still to be demonstrated:

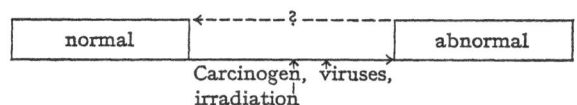

Fig. 1. See STOKER (1964), STOKER and McPHERSON (1961) and others: SACHS (1966)

There are signs that experimental tumours may originate by the trigger action of one causative factor. But, there are perhaps less dramatic indications that tumour formation rests on the right combination of multiple factors: carcinogens and co-carcinogens in mice, species specificity in animals, influence of sex and hormonal status, difference in epidemiology of certain forms of cancer in different nations and races (stomach cancer, breast cancer, etc.); potentiating effects of psychosomatic situations, age, familial disposition, etc.

II. General trends

Bird's eye view. When reviewing a given field of Medical Research, it is advantageous to take note of what has been said during conferences, discussion meetings, etc. over the last three or four years (true progress is usually not so rapid as to be restricted necessarily to a shorter period). It is from the papers and remarks made on such occasions (see MATHÉ, 1969a), that an overall picture of trends emerges which, if one searched for it in the numerous articles and publications in all the journals which deal with the subject of cancer, would appear as hundreds of fragments and not as a whole. Of course, even conferences do not disclose all the finer changes in working programs and events at the bench or in the clinic, that are taking place at this very moment or will take place during the next few months or years. But this is when one has to activate one's imagination and guessing talents. At the same time it ought to be understood, as mentioned at the start, that what follows in this review does not pretend to be complete. It is illustrative and critical rather than exhaustive.

Some recent international, national and regional meetings. An international meeting, organized by the Swiss Academy of Medical Sciences and sponsored by F. Hoffmann-La Roche of Basle in honour of their Research Director P. A. PLATTNER (1964) who acted also as editor, covered the now traditional subchapters of "cytostatic agents" (a name in use rather more in Central Europe than in other countries where "oncostatic or carcinostatic" appear more often in print), such as alkylating agents, antimetabolites, antibiotics, plant alkaloids, hormones, and a number of newer remedies such as a methylhydrazine (Natulan) and terephthalanilide derivatives. A thorough discussion of the effects of the older and newer compounds on neoplastic diseases, their selective toxicities, if any, and of the modes of action, especially of the alkylating agents, formed an important part of the proceedings which showed a happy mingling of clinicians and "backroom boys". Now, nearly five years after this conference, the terephthalanilides have practically disappeared from the scene because of their serious side effects which in analogue compounds did not substantially diminish; the methyl-hydrazine derivative and some of the antibiotic and alkaloidal agents, mentioned during the sessions, have taken their place in the armamentarium of cancer therapeutic compounds, side by side with older ones.

Brief mention should be made of a meeting in Cambridge, England, where a course was organized by J. S. MITCHELL (1965) on all "Aspects of Treatment of Cancer." By teaching post-graduate "students", a review of all the means of therapy of neoplastic diseases emerged, which in this case included radiotherapy in addition to chemotherapeutic techniques. While the experimental and clinical application of chemicals was surveyed on this occasion by the author of this review, a portion of the sessions was dedicated to radiation chemistry and biochemistry, underlining the close relation between physical and chemical approaches to treatments. In going over the various contributions, one can also see that a good deal of material of a purely clinical nature intermingled with more theoretical researches. Among newer developments at that time, one finds indications as to the possible role of immunological methods; this, in the light of recent advances in this special field, is historically interesting, as it mentions some work of HADDOW and ALEXANDER (1964) on rat tumours and their "immunological" treatment, a piece of research that represents the beginning of the investigations by ALEXANDER, DELORME and others (see later) into lymphocyte effects on tumour-bearing animals.

The next assembly, dedicated solely to carcinochemotherapy, that is worth while mentioning in this context took place in Osaka, Japan, following immediately the IXth International Cancer Congress in Tokyo (GOLDIN et al., 1967). It was sponsored by the Takeda Science Foundation and the Japanese Cancer Association. It aimed at an earnest investigation into the question, why a "breakthrough" of the formidable defences surrounding these fields of medical endeavour had not taken place and was not going to take place within the next few years. The answer may lie in the fact mentioned by the pastmaster of Japanese Cancer Research, T. YOSHIDA, namely that at present the guiding principle of the chemotherapist is more the strategy of containment than EHRLICH's "*therapia magna sterilisans*". However, a faint indication of things to come can be discerned in his words: it points to the need for a greater knowledge of regulatory mechanisms in contrast to the philosophy of complete eradication which is now predominant. One of the lines of research might be the intensification of studies of mechanisms of anti-tumour action, another, the exploration of improvements of testing methods (as long as one knows for what to test), and again another, attempts at solution of the intractable problems of selectivity and of resistance. The most pressing need is for the understanding of the relationship between results in animals and those obtained in man, and for the extension of treatment to include metastases (see GARATTINI et al., 1968b).

Three helpful summaries and statements of perspectives were offered by SHEAR, BURCHENAL and GOLDIN, all of whom pointed to collective and painstaking research, based on a much closer collaboration of clinical and scientific departments. Of course, as long as all these common-sense proposals remain pious wishes and verbal and written exchanges of views in committees, and do not turn into rather ruthless *practical* policies, they remain nicely printed parts of the conference procedures. It is strenuous uphill-work to achieve the transcription of certain ideas into actions which go somewhat outside the traditional activities of medical and scientific teams. The three "Wise Men", two experimentalists of yesterday and today and one clinical statesman, did express all this in a form dependent on their personal background.

As to the other reports and forecasts made at this conference, nearly a quarter of the talks was given over to the modes of action of carcinochemotherapeutic agents and the improvements of bioassay methods, nothing startlingly new, although the shift from novel remedies toward work on techniques and clinical administration is significant. At the time of this conference its atmosphere was polluted by the pessimism which had been spread in form of whispers in the corridors of the various hotels in Tokyo where the IXth International Cancer Congress did take place. This emergence of doubts about the present and future of 'Chemotherapy' as one of the acceptable means of treatment of cancers, was not shouted from the housetops as "it can't be done", "it's impossible," "it does more harm to the patient than good", but was expressed in subtle forms of shrugging shoulders, ironical smiles and overemphasis on the fundamental parts of research programmes, considering chemotherapeutic research as a very crude form of empiricism. To be fair, this was strongly counteracted during the symposium in Osaka, but everyone was somewhat on the defensive.

Apart from the Osaka conference and meeting of specialists all over Japan during the autum of 1966, a panel discussion had taken place on the subject of carcinochemotherapy while the International Congress was in session in Tokyo (see YOSHIDA, 1966; HARRIS, 1967). Unfortunately, the pastmasters in this field of cancer research, SIDNEY FARBER (1967) and L. LARIONOV (see 1966), were prevented from going to Japan; this deprived the special assembly of their counsel of long and outstanding experience. However, T. HALL (1966)[1] in place of FARBER made a strong plea for a more individualised treatment

---

1 No details of the Panel 4 discussions are given in HARRIS (1967).

of patients, based on the biochemical properties of the leukemia or tumour of a special cancer case, either in terms of its own metabolism or its specific reactions to drugs applied against it. Still within this Congress during sessions, when proffered papers were read under the heading of Experimental Chemotherapy, a number of claims for new agents were submitted of which only a few showed interesting features. Some of these agents will be mentioned later on, but items such as clam extracts, cytosine arabinoside, asparaginase and polycarbonyl compounds did serve as indicators of roads along which the fraternity of chemotherapists were and are marching.

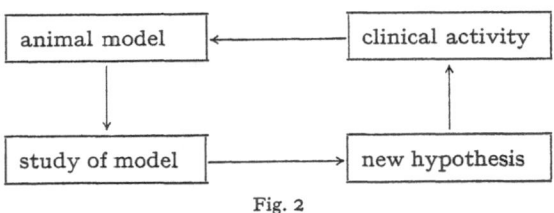

Fig. 2

During a symposium on Experimental Aspects of Cancer Chemotherapy at the Imperial Cancer Research Fund, London, organised by K. HELLMANN (1968), but not published, two philosophies of how to discover effective chemotherapeutic agents or to ascertain their desirable activities in man were propounded and discussed by two scientists of repute. One, T. CONNORS, like HALL in Tokyo, defended the thesis that new types of screening methods with a selection of new tumours—and human at that, in the form of hetero-transplants or cell or organ cultures, should be urgently developed. This is necessary for the following reasons: a) the rapid growth of transplanted tumours in animals, which is in complete contrast to the behaviour of human ones which grow relatively slowly; b) the resistance of human neoplasms vis-à-vis many agents, a phenomenon which cannot yet be explained satisfactorily; c) the considerable difference, either in animals or man, in the occurrence of various side-effects, including immune responses. This proposition contrasted sharply with that expressed by ZUBROD (1964, 1967), (ZUBROD et al., 1966; GOLDIN et al., 1966; LOO, 1966; RALL and HOMAN, 1967; DEWYS et al.,1968). The experience of the Cancer Chemotherapy National Service Center (CCNSC) and that of the National Cancer Institute since the re-organization of the Center has led them away from the vast and somewhat indiscriminate screening programme of former years toward a streamlined one, for which mainly GOLDIN and his colleagues on the experimental side and ZUBROD on the clinical side were responsible. They made use of the mouse leukemia L 1210 to discover whether—in the first place—older drugs, when "chased" through an experimental cycle like the one depicted in Fig. 2, would show expected or unexpected effects. They also argued that primary screening-predictions were in the past mainly based on fast growing abnormal tissues. They asked themselves what differences existed between one or the other type of cell population (Fig. 3).

The answer was: the difference lies in the size of the proliferating pool and the relation of the rate of growth versus the rate of cell death. These are kinetic phenomena. Thus the NIH screening rests on a basis of cell kinetics and a one-tumour system.

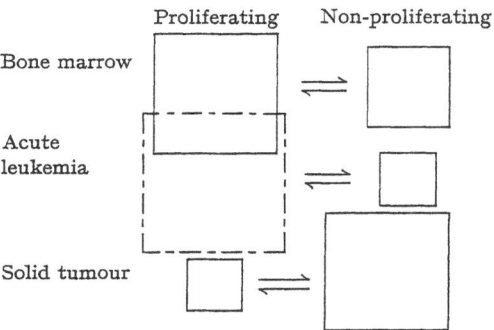

Fig. 3. The squares indicate proportions of cells however representing no absolute but only relative values. With acknowledgements to Dr. C. G. ZUBROD

Some of these debates and arguments have turned up during a threestep sequence of Symposia centred on leukemias and related diseases (MILDER, 1967). They formed part of a symposium on The Rationale and Assessment of Cancer Chemotherapy in April 1969 in Manchester, England, under the auspices of the British Association for Cancer Research and at a Conference on The Critical Evaluation of Cancer Chemotherapy held at the same time in Cherry Hill, New Jersey; the latter was organized by a Committee of the Union International Contre le Cancer (UICC) and the American Cancer Society (GARATTINI et al., 1969; MANDEL, 1969). Other portents of change of ideas and working programmes can be gleaned from proceedings of meetings which have taken place during the last 12 months, very often on specific subjects, such as L-asparaginase. This topic was thoroughly illuminated from different angles during a conference at the Squibb Institute for Medical Research, New Brunswick, in March, 1968. On this occasion BROOME (1963, 1966, 1968a) continued his story of this enzyme which he had reported in part during the Tokyo Congress. A few weeks after the Squibb Conference the American Association of Cancer Research devoted one of its sessions in Atlantic City to the same subject matter. The emphasis there was on the possibility that asparagine synthetase played an even bigger role than the degrading enzyme (HOROWITZ et al., 1968). This will be considered further on.

During the Meeting in Atlantic City (SHIMKIN, 1968) a symposium on Immunological Aspects of Malignant Diseases was arranged to explore principally the background of this new (or better *renewed*) branch of carcinochemotherapy, in the widest sense (PRESSMAN, 1968). It is worth while remembering that one of the conferences on Acute Leukemia and Burkitt's Tumor dedicated one of its sessions under C. M. SOUTHAM (1967) to immunological problems of these malignant states. This indicates again the increasing attention which the cancer investigators are paying now to this topic. It appears justified to consider it under the heading of chemotherapy, as in the long run it will become clear that it is the effectiveness of the molecular and consequently chemical components of the immunological process which counts. This even extends to cell-bound constituents, a matter which will be explained, when the progress in the carcino-immunological field is scrutinized. Coming back to the Atlantic City Meeting of Spring 1968, the Presidential Address, delivered by L. W. LAW on this occasion, had as its title: Studies of the Significance of Tumor Antigens in the Induction and Regression of Neoplastic Diseases (LAW, 1969); and P. GOLD (1968) as a participant of the symposium developed very positive ideas on the basis of encouraging observations in patients with cancers of the digestive system. At a recent meeting of the Federation of American Societies for Experimental Biology (1969) KLEIN gave an address which produced the keynote to the Cancer section. He underlined that, at least in some cancers, there operated an immune system which was capable of controlling constantly the genesis of small numbers of cancer cells.

While papers bearing on the subject of carcinochemotherapeutic problems can be found scattered throughout meetings, such as those held during fall of 1968 and spring of 1969 by the British Association of Cancer Research and similar national organizations, the Committee on Experimental Chemotherapy of the U.I.C.C. Commission on Experimental Oncology under GARATTINI, CONNORS and ALBERT (1968a) arranged in October, 1968, a special training course in cancer chemotherapy in Warsaw, Poland. This meeting attempted a survey of the practical issues at present in the forefront of interests of oncologists, such as cellular structure of tumour cells, classification of tumours, the use of transplanted tumours and, last but not least, the question of tumour resistance. One interesting item was an attempt to compare the screening methods as used in a number of places, such as the NIH, Geca[2],U.S.S.R. Academy of Medical Sciences, etc. Another was the present state of *in vitro* tests (see also BERGEL, 1966b), the problem of metastases, drug metabolism, combination therapies, activation and deactivation of anti-cancer agents and, as pointed out above with regard to the topical position of carcino-immunological problems, an assessment of this revitalized research chapter.

---

2 Now OERTC = European Organization for Research into Cancer Treatment (MATHÉ, 1969a).

III. Has true progress been made?

A. The concept of eradication

After having taken a look at the trends, developed or developing, it is necessary to produce more detailed information about the various groups of remedies so that the rough sketch, so far drawn, can slowly transform itself into a more coherent picture.

1. Alkylating agents

Comprehensive reviews for obtaining a good background are, for instance, the monograph by Ross (1962) and among many others a treatise in form of a chapter in Metabolic Inhibitors written by JOHNSON and BERGEL (1963).

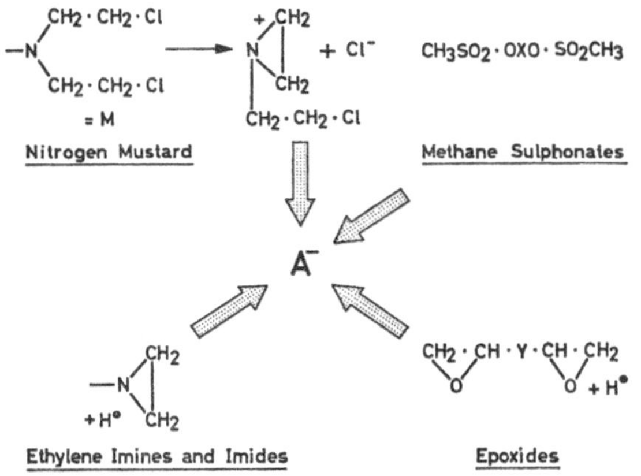

$$CH_2 \cdot CH_2 \cdot Cl$$
$$-N \qquad \longrightarrow \qquad N^+ \begin{matrix} CH_2 \\ | \\ CH_2 \end{matrix} + Cl^- \qquad CH_3SO_2 \cdot OXO \cdot SO_2CH_3$$
$$CH_2 \cdot CH_2 \cdot Cl \qquad \qquad CH_2 \cdot CH_2 \cdot Cl$$
$$= M$$

Nitrogen Mustard   Methane Sulphonates

$$A^-$$

$$-N \begin{matrix} CH_2 \\ | \\ CH_2 \end{matrix} \qquad\qquad CH_2 \cdot CH \cdot Y \cdot CH \cdot CH_2$$
$$+H^\bullet \qquad\qquad\qquad O \qquad\qquad O + H^\bullet$$

Ethylene Imines and Imides   Epoxides

Fig. 4. Prototypes of alkylating groups. (From BERGEL, 1964a.) With grateful acknowledgements to P. A. PLATTNER (Editor), Chemotherapy of Cancer, Elsevier Publishing Co., Amsterdam, 1964

The Ross monograph—in addition to other information—still represents the best basis for all the theoretical problems connected with this species of anti-cancer agents. A further review of alkylating agents up to 1963 is available in Cancer Chemotherapy Reports (BRATZEL et al., 1963) and a survey of the comparative pharmacology of these compounds in the same journal (SCHMIDT et al., 1965) in the form of a mammoth collection of data. A later review has appeared in volume V of Experimental Chemotherapy covering the middle sixties (OCHOA and HIRSCHBERG, 1967).

As pointed out on a former occasion, the alkylating agents (Fig. 4) are not more than the bows and arrows in the armamentarium of anti-cancer agents (BERGEL, 1964a). They are effective in killing fast growing cancer cells and in arresting mitotic activities. They can be used profitably as research tools but should be applied only with discretion and caution in therapy.

The shortest way to remind the reader of the state of affairs, as it prevailed up to a few years ago and may continue to represent the therapeutic situation

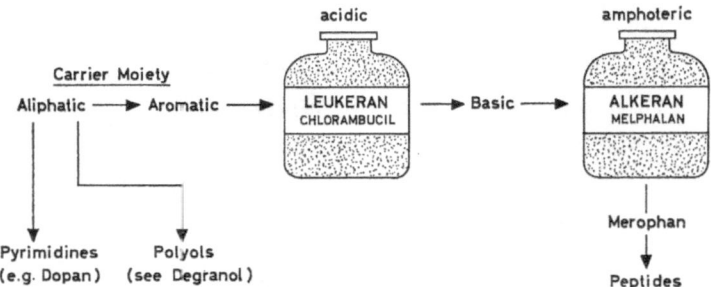

Fig. 5. Development of N-mustards I. (From Bergel, 1964a.) With grateful acknowledgements to P. A. Plattner (Editor), Chemotherapy of Cancer, Elsevier Publishing Co., Amsterdam, 1964

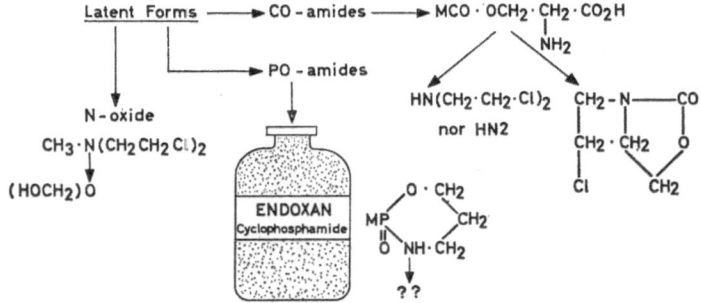

Fig. 6. Development of N-mustards II. (From Bergel, 1964a.) With grateful acknowledgements to P. A. Plattner (Editor), Chemotherapy of Cancer, Elsevier Publishing Co., Amsterdam, 1964

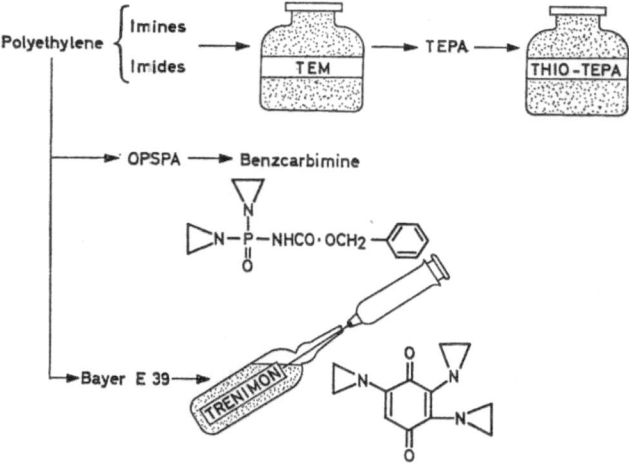

Fig. 7. Development of methane sulphonates and epoxides. (From Bergel, 1964a.) With grateful acknowledgements to P. A. Plattner (Editor), Chemotherapy of Cancer, Elsevier Publishing Co., Amsterdam, 1964

for quite a few years to come, is to borrow a set of figures (Figs. 5, 6, 7 and 8) from a paper already referred to by the reviewer (Bergel, 1964a).

The bottles or ampoules indicate the clinical use, based on commercial availability. Going back to the same period, i.e. about five years ago, one

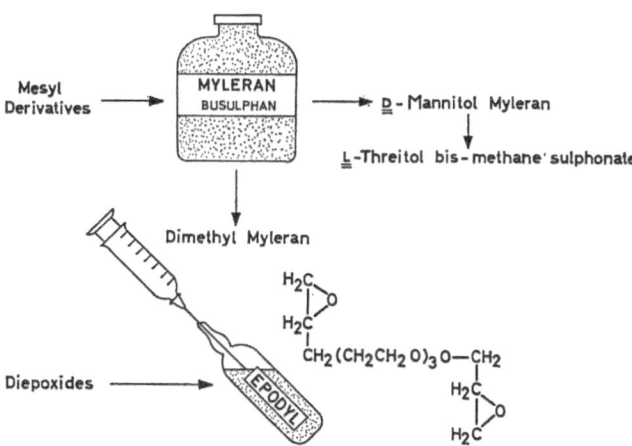

Fig. 8. Development of ethylene imines and imides. (From BERGEL, 1964a.) With grateful acknowledgements to P. A. PLATTNER (Editor), Chemotherapy of Cancer, Elsevier Publishing Co., Amsterdam, 1964

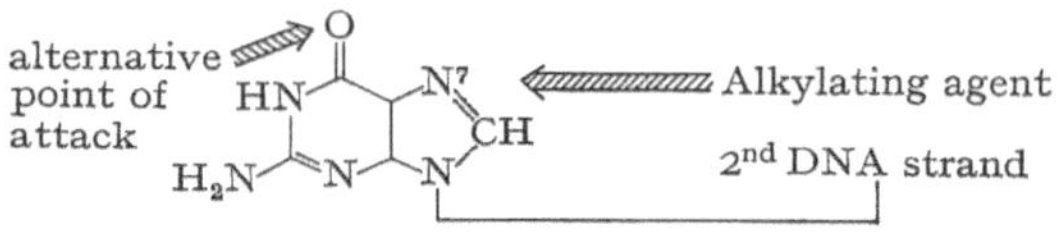

Fig. 9

finds with reference to the mechanisms of action the impressive experimental observations of LAWLEY and BROOKES (see BROOKES, 1964), now generally accepted, that the therapeutic and carcinogenic effects of these compounds are due to the interaction of the alkylating agents with the N^7 of the guanine moiety of DNA, with a crosslink to the second DNA strand (Fig. 9).

Ever since LAWLEY (1966) published his original ideas, a controversy has been going on concerning the following questions. Is the "Lawley interaction" the true and sole mechanism of the chemotherapeutic and/or carcinogenic effects? Are the real causative events due to a multimechanistic effect, in the sense that several interactions of the cytotoxic agents with a number of diverse cellular components have to happen synchronously or sequentially, the components being, in addition to DNA, proteins and in particular functional proteins, for instance NAD-pyrophosphorylase, as proposed by HOLZER (1961, 1963 and 1964)? Even taking the guanine moiety of DNA as the principal target of the alkylating attack, the N^7 (Fig. 9) might not necessarily be the only vulnerable point of the purine moiety. According to LOVELESS (1969), alkylating agents of the N-alkyl-N-nitroso-urea type, $ON.N(CONH_2)CH_3$ or C_2H_5 (see DRUCKREY et al., 1967), with predominantly mutagenic action on T_2 bacteriophage may prefer to assault the oxygen of the pyrimidine part of, say, deoxyguanosine. In consequence, it seems that the underlying mechanism of alkylation damage to cells, comprising mutagenesis, carcinogenesis or lethal effects, is not yet known in its entirety.

In the following, the progress of preparation, application and testing of newer agents will be briefly considered. It should be obvious that only a selection of examples can be given and a hint or two as to where one can find others.

In this reviewer's opinion, one of the most interesting observations concerns the simplest aromatic mustard, namely di-(2-chloroethyl) aniline or aniline mustard. CONNORS and WHISSON (1966) and their collaborators discovered that this uncomplicated alkylating compound possessed an outstanding activity against a mouse plasma-cell tumour which shows some similarities to the human neoplastic disease of multiple myeloma. Not only did aniline mustard cure completely large experimental tumours when administered some time after implantation, but it showed relatively little toxic effects in the doses used, in contradistinction to the p-hydroxyderivative. Other para-substituted derivatives proved to be much less, if at all, carcinolytic. As one of the modes of action, it was proposed that the original compound was transformed in the liver into the glucuronide of the p-hydroxyaniline mustard and broken down selectively in the tumour to the phenolic material by glucuronidase which is present at a high level in the plasma-cell tumour used (ADJ/PC5) (ROSENOER and WHISSON, 1966). Further experiments with the labelled agent ([³H] aniline mustard) disclosed high levels of activity remaining in the bile and serum of the mouse. A search for patients with myelomatosis who may react equally favourably to the drug, because of high activity levels of the catabolic enzyme in the neoplastic tissue, is still going on.

Another piece of experimental work concerns the systematic investigations carried out by Russian (LARIONOV, 1965) and British (BERGEL, 1964a) workers on a series of aromatic amino-acid derivatives. It started with the phenylalanine derivative melphalan (sarcolysine), continued with di- and polypeptides (BERGEL et al., 1965; SZEKERKE et al., 1968) that showed interesting stereospecificity of activity (BERGEL, 1964b), and finished, so far, with the preparation of protein derivatives (WADE et al., 1967). In the latter series, which merits further research, it was noteworthy how much the therapeutic index (LD_{50}/ID_{90} = Th I) vis-à-vis the mouse plasma cell-tumour ADJ/PC6A, similar to that mentioned above, had improved, if aromatic nitrogen mustards carried fibrinogen or albumin or if they even were only mixed with these proteins. On the average, the index rose by factors of 6—120 under the experimental conditions of the tests.

In Fig. 7, the methane sulphonate derivate busulfan (Myleran) and a poly-hydroxy analogue, mannitol-myleran, were mentioned. Recently, ELSON, JARMAN and Ross (1968) quoted interesting results that relate dianhydromannitol to mannitol-myleran and the corresponding 1,6-dibromo-1,6-dideoxy-D-mannitol (DBM) (HORVATH and INSTITORIS, 1967). The conclusions are that mannitol myleran and DBM may produce their chemotherapeutic effects by being converted *in vivo* into the dianhydro-derivatives. This means that it is the di-epoxides of this series which exert the alkylating effects and not the mesyl compounds or the di-bromides (see also JARMAN and ROSS, 1969a). The animal tests which formed the basis of these observations were carried out on the ADJ/PC6A mouse tumours, on the WALKER 256 rat carcinoma, and on the L 1210 mouse leukemia. Another methane sulphonate has had experimental success; it is a derivative of formaldehyde, methylene dimethane sulphonate (FOX, 1969).

It is an old aim of carcinochemotherapy to increase the selectivity of agents such that they act in the main on those parts of organs and tissues that show large areas of neoplastic changes. This can happen through metabolic events, as attempted with aniline mustard and mentioned previously, or by providing a molecular locking device that will be removed in and by the malignant tissue (latent agents), or by carriers which facilitate the transport of the chemotherapeutically active compound to the tumour (see the references to the amino acid derivatives). If it is correct to assume that at least certain cancer cells have a pH of their cytoplasmic contents which is lower than normal, i.e. on the acidic side, then any compound, that at a physiological pH (about 7.2) possesses a moderate solubility, would be precipitated under conditions of lower pH. CALVERT, CONNORS and ROSS (1968) have found experimentally that a number of aromatic nitrogen mustards carrying sulphone amide groups fulfill these expectations, and one of them, namely the propyl analogue of a sulphadiazine derivative, proved to be exceptionally effective against Yoshida and Walker tumours and showed a greater concentration in the tumour than in the liver, a selectivity which is in contrast to that of other aromatic alkylating agents. While clinical proof that these experimental results will hold good under human conditions has still to come, it is interesting to note that a number (small it must be confessed) of rational ideas do pay off, even in this tricky field of alkylating agents.

If somebody surveys the literature of today in order to find novel and exciting alkylating agents, he will be on the whole rather disappointed. Most of the clinicians engaged on a national level in therapeutic practices have their drugs of preference, which are locally manufactured and with which they continue to treat leukemias and solid tumours. For instance, cyclophosphamide (Endoxan, Endoxana) has not only maintained its position but gained ground in the fields of leukemia, Burkitt's tumour and among immune suppressors (see HAMILTON FAIRLEY and SIMISTER, 1964 and later literature). Trenimon has still its loyal users (HOELZEL et al., 1968; LISS et al., 1968) and Natulan is being used in Central Europe (WEITZEL et al., 1968). These regional predilections underline the small quantitative differences in curative powers between one or the other cytotoxic drug, except in cases of certain cancers. Adopting a pragmatic stance one cannot quarrel with this attitude, particularly if it does not resist fresh ideas about dosage, duration of treatment, or time intervals of application, and about combination with other drugs. FARBER (1965), during one of the leukemia conferences, pointed out that clinical research in depth was necessary, relatively much more than a search for new remedies. Consulting the Abstracts of the IXth International Cancer Congress (1966), one finds there a Czech report on thymine mustards (VESELY, 1966), a piece of work which follows in the footsteps of the workers at the USSR Academy of Medical Sciences, foremost LARIONOV et al. (1965). It represents an example of the carrier principle. A remarkable improvement of a monofunctional alkylating drug has been achieved by COBB et al. (1969) in the form of a dinitro-ethyleneiminobenzamide with a specifically high activity against the Walker carcinoma; if an equivalent of the rat tumour is found in man, this compound might well serve as an object of profitable clinical research. Otherwise it is surprising to discover how many laboratories manage to produce a considerable number of certain alkylating agents and claim remarkable activities; all this with the use of rather doubtful screening procedures, and with the whole project from the beginning based on a weak rationale. This is reminiscent of the complaints raised on many occasions against a number of industrial laboratories which, so it is believed, play a kind of drug roulette by producing variants of a successful prototype, without achieving any advance on the first drug on the market (often possessing specific pharmacodynamic properties). But this happens also with academic and specialist institutes who join the race for the high price of a maximally effective anticancer agent by repeating the synthesis of closely related compounds showing only small differences in structure and therapeutic properties. Unfortunately, one cannot disclose the names of these laboratories and workers without having tested these innumerable drugs in comparison with generally accepted standards. However, one must allow for the possibility that among the many agents there might be unexpectedly one which shows much greater promise than any of the others. Nevertheless, as was pointed out previously, it can be said with certainty that, in the field of alkylating agents, the difference between one and the other compound as to its clinical efficacy is relatively small and can only be expressed in small multiples of activities. Thus further expenditure in work and money should be carefully re-directed into other fields which have been ploughed and reploughed less often.

2. Antimetabolites

In order to find out most of this group's background, it suffices to browse through an article by STOCK (1966a) and a monograph, "The Antimetabolite Approach" by TIMMIS and WILLIAMS (1967). The latter was published 20 years after FARBER and his colleagues reported their findings with a folic acid antagonist in the treatment of leukemia in young children. Most of the drugs which followed this basic and most important discovery are still in use and the major progress has been more through building on the knowledge of these first years of success than finding strikingly new routes of development. The only exception may be nucleosidic derivative compounds. On the whole, experimental and clinical workers in carcinochemotherapy still deal with antipurines, antipyrimidines, antifolics and a number of other chemicals which, following a kind of allosteric mechanism, act, in mammals, either directly or indirectly (see MONOD et al., 1963) on essential metabolites among which amino acid derivatives and vitamins play the most important part.

Since aminopterin was the first effective chemotherapeutic agent, a quick look at other folic acid antagonists should give one an idea into which direction investigators in this special field are moving. As TIMMIS and WILLIAMS' book gives mostly the chemical mechanistic background, a volume, edited by PORTER and WILTSHAW (1962) and forming a symposium report on "Methotrexate in the Treatment of Cancer," is very helpful. Following a historical introduction by one of the discoverers of the antagonist drug, T. H. JUKES, a comprehensive survey of the biochemistry by BOYLAND and of the pharmacology by FREEMAN-NARROD, a galaxy of clinicians described in detail their personal experience of various types of cancer, with the greatest interest centered on chorionepithelioma, one of the few tumours which respond successfully to chemotherapy (HERTZ et al., 1961; ROSS et al., 1962); details of the techniques used in intraarterial therapy and intrathecal administration of this relatively versatile drug were also presented. At the end of three clinical sessions, the chairman underlined the importance of intensified studies of carcinotherapeutic agents, even if they had been in clinical use for some years; such studies are and should be directed towards the application of new techniques of administration, the accurate establishment of the circulating drug level and of the rate of excretion, and to the ease of transport of the chemical to the sites of action. These sites were in many cases the head and neck regions, the treatment of which with methotrexate, and the use of leucovorin as an antidote, has been reported more recently by LEFKOWITZ et al. (1967). Of course, reading about these clinical investigations, of which this is only an example, one is sadly aware that the requests of some researchers such as T. C. HALL (1966) or T. A. CONNORS (1968) are still largely ignored, because the assessment of the serum levels and excretion rates of the drug is no substitute for a thorough biochemical investigation of each patient as to his or her specific drug metabolism and resistance (see ROBERTS and HALL, 1969).

Going over the subgroup of anti-purine and anti-pyrimidine derivatives, one is struck by the increase, during the last few years, in the use and importance of nucleosides, originating with normal nucleic acid bases combined with unnatural sugars, and of those which represent sugar derivatives of analogues.

Table 2. Nucleoside derivatives of clinical interest

NSC No.	Name	Group No.	Group name	Biologic activity*						Clinical activity
				L 1210 ILS	Walker TWI	Ca 755 TWI	S 180 TWI	Lewis lung TWI	KB ED$_{50}$	
27640	5-FUDR (2'-Deoxy-5-fluorouridine)	2	Uracil 2'-deoxyribosides	55	71	85	95	85	<1.0	+
38297	5-BUDR (2'-Deoxy-5-bromouridine)	2	Uracil 2'-deoxyribosides	22	55	62	33	59	30	±
39661	5-IUDR (2'-Deoxy-5-iodouridine)	2	Uracil 2'-deoxyribosides	40	72	60	50		14	±
75520	F$_3$TDR (2'-Deoxy-5-trifluoromethyluridine)	2	Uracil 2'-deoxyribosides	75	72	73	48			P
63878	Ara-C (Cytosine 1-β-D-arabinofuranoside)	6	Cytosine furanosides	133	32	86	85	45	<1.0	+
93150	2', 3', 5'-Triacetate Ara-C	6	Cytosine furanosides	90	14		68		<1.0	P
529180	5-Fluoro-Ara-C	6	Cytosine furanosides	156						+
4911	6-MPR (6-Mercaptopurine 9-β-D-ribofuranoside)	14	6-MP riboside analogs	55	87	96	81		<1.0	+
40774	6-Methyl-MPR	14	6-MP riboside analogs	60	23	68	28		<1.0	+
406021	6-MPAra (6-Mercaptopurine 9-β-D-arabinofuranoside)	15	6-MP (nonribose) sugar analogs	63		100		28	<100	P

NSC No.	Compound	Class							ED50	ILS
30605	2-Fluoro-AdR (Adenosine)	Adenine ribosides	19	12	55	13			<1.0	−
3055	Puromycin	Adenine ribosides	19	17	69	34			<1.0	−
29422	6-TGR (6-thioguanine 9-β-D-ribofuranoside)	6-Thioguanine ribosides		64	98	85	70		<1.0	+
71261	2'-Deoxy-6-TGR	6-Thioguanine (nonribose) nucleosides	31	99	98				<1.0	P
71851	2'-Deoxy-6-thioguanine 9-α-D-ribofuranoside	6-Thioguanine (nonribose) nucleosides	31	95	91				<1.0	P
56408	Tubercidin	7-Deazapurine ribofuranosides	34	16	25	30	10	40	<1.0	±
63701	Toyocamycin	7-Deazapurine ribofuranosides	34	25		28	21	17	<1.0	− (toxic)
65346	Sangivamycin	7-Deazapurine ribofuranosides	34	50		50	13	20	<1.0	P
32074	6-Azauridine	Miscellaneous nucleosides	36	35	35	0	43	37	<1.0	±
67239	2', 3', 5'-Triacetate 6-Azauridine	Miscellaneous nucleosides	36	44		57	26	11		±
102816	5-Azacytidine	Miscellaneous nucleosides	36	156	68					P

* Underlining denotes active ratings.

NSC No. = Number of sample of National Service Center, Bethesda Md.

ILS: increased lifespan. TWI: tumor weight inhibition. +: clinically active. ±: clinical activity moderate. −: clinical activity not demonstrated. Either inactive or too toxic. P: being actively processed for the clinic. NT: not tested. KB: Human epidermoid carcinoma of nasopharynx; figures refer to ED_{50} = μg/ml.

With grateful acknowledgements to Messrs. GOLDIN et al. and the Editors of the Cancer Chemother. Rep. vol. 1, No. 1, part 2 (Suppl.), 1—272 (1968).

The most comprehensive survey on the relation of structure to anti-tumour activity (test systems L 1210, Walker 256 carcinoma, adenocarcinoma 755, sarcoma 180 and the Lewis lung tumour in mice, and a tissue culture screen) were reviewed by GOLDIN et al. (1968). In all, 648 compounds were recorded belonging to 36 structural groups. To improve the usefulness of presentation, Table 2 is added; it gives the clinical activities so far studied, with negative and positive results. There are 21 nucleosides of clinical interest which belong to some, but not all, of the 36 structural groups. (Table 2 was made up from the original protocol.)

The one agent most frequently used in medical practice since its discovery in 1964 is the cytosine arabinoside (HOWARD et al., 1966; KLINE, 1966; ZUBROD, 1967). The use of this pyrimidine antagonist, perhaps also adenine arabinoside and the corresponding 5-fluorouridine (HEIDELBERGER, 1966) in a number of neoplastic diseases underlines the well-known fact that the corresponding nucleotides, that is the phosphorylated nucleosides, are therapeutically useless, mainly due to the charge effect of the phosphate group which hinders penetration of the cell membrane.

This problem was particularly studied in a group of purine antagonists by members of the Kettering-Meyer Institute (BROCKMAN, 1965; see TIMMIS and WILLIAMS, 1967). The working hypothesis in these cases was that the resistant tumours had lost the capacity to form nucleotides from the antagonist bases or their nucleosides. However, it has to be stated that resistance to anti-leukemic or anti-tumour agents may rest also on other types of mechanisms. One of the most interesting attempts to circumvent the obstacle of cell membranes to permeation by phosphates (when given as drugs) was to synthesize a compound with lower ionization properties than the monophosphate (THOMAS and MONTGOMERY, 1962). This was found in bis(thioinosine)-5',5'''-phosphate. However, comparable attempts with 6-azauridine to circumvent the inactivity of its nucleotide by transforming it into a double molecule, bis-6-azauridine-5',5'''-phosphate, ended with negative results (HANDSCHUHMACHER, 1965). Most of these difficult investigations were carried out with cells of animal tumours, and it remains to be seen whether corresponding resistance of human tumour or leukemia cells are due to similar shortcomings in phosphorylation mechanisms. Whatever the findings, it seems that the level of nucleotide formation, the determinant step in rodents, is not very high in most human leukemias (HALL et al., 1968; KESSEL, 1969).

To swing over from resistance to enhancement of activity, or perhaps more correctly to drug-sparing action, it is worthwhile to be reminded by the findings of ELION et al. (1963; see also 1966) of the potentiation of mercaptopurine and some of its derivatives by the xanthine oxidase inhibitor 4-hydroxypyrazolo- (3,4-α) pyrimidine (allopurinol), which has turned out to be an effective remedy against gout. This is due to the inhibition of the oxidation of these agents into uric and 6-thiouric acid derivatives. To complete this short list of anti-purine and anti-pyrimidine bases and nucleosides, it should be mentioned that N^6-(Δ^2-isopentenyl) adenosine has found some admirers and that adenine 1-N-oxide has inhibitory effects on experimental tumours in contrast to the N-oxides of guanine and xanthine and to 7-hydroxyxanthine (SUGUIRA and BROWN, 1966 and 1967).

With allopurinol one enters the field of specific enzyme inhibitors (TELLER et al., 1968) which might be considered as belonging to the larger family of antimetabolites. While the anti-neoplastic activity of L-asparaginase will be mentioned later on, asparagine antagonists such as aspartic acid hydrazide (OHNUMA et al., 1967) and even S-carbamoyl cysteine (ADAMSON et al., 1968) show some effects. The idea to look for asparagine synthetase antagonists would be rewarding only if the mode of action of asparaginase rested solely on deficiencies in certain lymphomas and leukemias of a single enzyme responsible for the biosynthesis of the amino acid amide; the possibility of the existence of several biosynthetic mechanisms cannot be excluded (BROOME, 1968a and b, 1969).

Among amino acid antagonists the most remarkable remains 1-amino cyclopentane-1-carboxylic acid (CONNORS et al., 1960) although inconclusive reports appear here and there of other analogues of this group of cellular components. But this kind of work can be considered rather out of date and, unless peptides with carcinolytic activities turn up among the antibiotics (see further on), this part of the field of antimetabolites will lose attraction for the research worker.

Interesting from a biochemical point of view, but not, as yet, in regard to practical application, are the researches in neighbouring fields of vitamin analogues (which includes, of course, antifolics), i.e. riboflavin-, nicotinamide-, or B_{12}-antagonists. On the whole, these compounds have not contributed notably to the number of effective chemotherapeutic agents. They show some antimicrobial effects but their anti-tumour activity is very limited. This would not matter at all, if a mammalian tumour would be discovered one day that depends on one or the other of these "growth factors". There are laboratories which, making use of a pause of over ten years, are restarting the attack from this angle with fresh vigour and new ideas.

One of the vitamins, nicotinamide, acting like others do as precursor of coenzymes, has been a target for a number of workers; this is because of its major role in the glycolytic pathway of energy production which, as WARBURG discovered, is prevalent in a majority of tumours. With the WARBURG concept this review enters the region of metabolic regulation which might lend itself to interruption and inhibition by coenzyme analogues. Of the older compounds, 6-aminonicotinamide, and thiadiazole derivatives are interesting enough to be mentioned again [see TIMMIS and WILLIAMS, 1967

(p. 171)]. More recently Ross (1964), a strong believer in the lactic acid specificity of tumour tissue, has undertaken the synthesis of a number of nicotinamides (Ross, 1966, 1967a and b; LOVESEY and Ross, 1969) under the heading of anti-coenzymes; they should form NAD-like compounds which might act as fraudulent coenzymes, interfering with normal functions. In addition, nicotinamide ribosides have been obtained by Ross and JARMAN (1966, 1967) and by JARMAN and Ross (1969b). At present the compounds exert relatively small chemotherapeutic activities. But the underlying rationale of their role is important for future developments. As can be gathered from Fig. 10 hydrogen transfer by NAD is stereospecific. The hydrogen atom is accepted in beta-position (Fig. 10a). In the subsequent reduction of pyruvate to lactate, NAD is regenerated from NADH by the transfer of the alpha-hydrogen atom (Fig. 10b). In the reactions c and d of a substituted nicotin-amide, c can proceed but d (the reduction of pyruvate to lactate) is blocked.

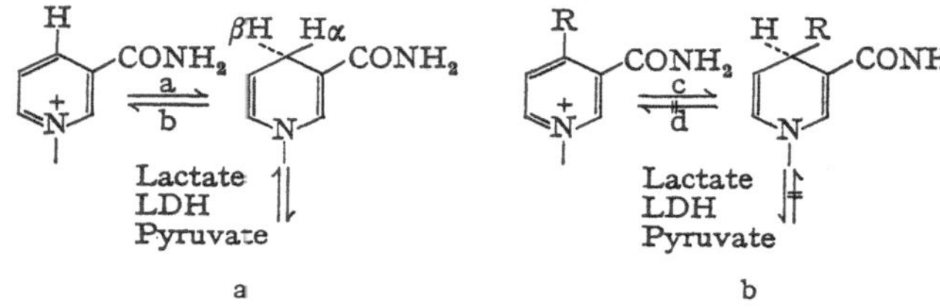

Fig. 10. With grateful acknowledgements to W. C. J. Ross (1964) and the Editors of the Annual Reports of the British Empire Cancer Campaign

B. Catabolic, metabolic substances and natural products
1. Enzymes and related carcinostatic agents

It is appropriate to continue this review with some remarks on enzymes, particularly as these follow the paragraph on potential anti-coenzymes. Some years ago, this reviewer wrote a small monograph on the potential role of enzymes in cancer research (BERGEL, 1961). It will be remembered that, among the experimental examples given, most of the biocatalysts were part of nucleic acid metabolism or at least involved in some control of purine cata-bolism. As it turned out, the enzyme which more recently caught the attention of a good number of workers was one that influenced greatly the levels of asparagine in biological systems, namely asparaginase. KIDD (see BROOME, 1961) was the first to report inhibition of a number of mice and rat lymphomas by guinea-pig serum; however, BROOME (1961, 1963, 1965, 1966, 1968a and b) was the discoverer of the fact that the biological activity of the serum was due to the presence of asparaginase. The increasing interest produced a number of different ways of expanding the work. OHNUMA et al. (1966) investigated chicken liver and WASHBURN and WRISTON (1964) *Escherichia coli* as sources of enzymes with somewhat different types of activities. ADAMSON and FABRO (1968) published a critical review of the biological properties of asparaginase. Meetings during the same year, at the Squibb Institute for Medical Research (1968), and at the Annual Meeting of the American Association for Cancer Research (SHIMKIN, 1968), dealt with recent biochemical and mechanistic points of this enzymotherapeutic problem. In both instances a number of clinical workers reported also on their experience with purified enzyme, mainly of bacterial origin, in cases of leukemias and lymphomas (see also HILL et al., 1967 and OETTGEN, 1967). It will require at least two or three years of clinical research before a final conclusion can be drawn as to the utility of the preparations, produced on a large scale, against various types of diseases of the blood and the lymphatic system, including Burkitt's variant (see KEAST et al., 1968; they used guinea pig serum against reovirus-induced mouse lymphoma 2731/L which appears to be related to Burkitt's lymphoma). The other task which needs further efforts is the elucidation of the underlying mode of action of the enzyme linked with the role that asparagine synthetase may play in obtaining positive or negative therapeutic results (see BROOME, 1969; LAZARUS et al., 1969).

It is appropriate to remind the reader here of the possibility of finding carcinotherapeutic agents, not so much among enzymes with all their disadvantageous properties of instability, scarcity and antigenicity, but among agents which possess enzyme-like properties and which one could call enzymomimetics (BERGEL and HARRAP, 1965). The basic idea has been expressed in the monograph mentioned before (BERGEL, 1961) and on several other occasions. The practical outcome depends on these two conditions: the first is to find a combination of a co-enzymal compound with activators which produces the same catalytic effects that are produced by the enzyme to be imitated and which are desirable from a therapeutic point of view. (The V_{max} of the enzyme model—substrate interaction will be of a much lower order than that of the biocatalyst, but this is of no imporance as long as it possesses sufficient specificity.) The second condition refers to the fact that coenzymes are frequently phosphorylated vitamin derivatives which may have difficulties in penetrating cellular membranes. This means that the enzymomimetic compounds should be capable of being phosphorylated *inside* the tumour cell or should be phosphates which have low ionization (see THOMAS and MONTGOMERY, 1962). In an example which utilizes pyridoxal-phosphate and vanadium salts as a model for cysteine desulphydrase, the reviewer and his colleagues (BERGEL et al., 1962) some years ago thought that deprivation of the cancer cell of cysteine, as of asparagine, might produce anti-tumour or anti-leukemic effects. Recently FOLEY's team in Boston (FOLEY et al., 1969) demonstrated the dependence on cystine of human leukemic lymphocytes in cell culture. Coming back to the investigations in

England, when substituting the pyridoxal phosphate with vitamin B_6, pyridoxine, no convincing clinical effects were produced, such as reduction in the number of specific leukocytes, because — as discovered later — patients, for instance, suffering from granulocytic leukemia are less able than normal people to transform pyridoxine into pyridoxal phosphate (BERGEL and HARAPP, 1965); [this is somewhat reminiscent of the mutants of experimental leukemias and their deficiency in ribophosphorylating enzymes for the intracellular transformation of antipurines into antipurine nucleotides (BROCKMAN, 1965)]. In spite of this apparent failure in therapeutic success in this case, it is believed that further work in the region of enzymomimetic compounds or similar combinations is very desirable. In some respects they might have to be considered as potential regulators, to be discussed below.

This is the best place to mention a number of agents which carry keto- or aldehydo-groups, or belong to the group of polyamines (of which several occur naturally in mammals), or are derivatives of metabolic products, such as urea, because they are in many instances acting as substrates or forming products of enzymes, inside metabolic pathways.

To start with, glyceraldehyde, which in its phosphorylated form participates in glycolytic energy metabolism, has been reported together with similar compounds as possessing anti-tumour effects (WARBURG et al., 1963; GERICKE, 1967a; LEWERENZ, 1967). However, possibly through its instability these effects are not very pronounced. During the last few years, some tentative reports were received from Germany about the clinical use of methylglyoxal, another three-carbon ketoaldehyde. This compound also turned up in connection with the work of SZENT-GYÖRGYI (1968) in his search for the elusive promin and retin couple, claimed to exert a regulating effect in various tissues. Several research workers (FRENCH and FREEDLANDER, 1960; BARRY et al., 1966; REGELSON et al., 1967; see also STOCK, 1967) have tested the idea that aldehydo- and keto-derivatives in form of thiosemicarbazones, or bisguanyl-hydrazones etc. could act as anti-tumour agents. But none of them has developed beyond the experimental stage (MIHICH and HAKALA, 1968).

One other type of potential anti-tumour agent is worth adding to this subchapter, namely simple hydroxylamine derivatives (see STOCK, 1967), such as N-hydroxyurea (D. T. KAUNG et al., 1968) and hydroxyurethane (KRAKOFF and CLIFFORD, 1968). Both appear to exert qualitatively similar effects but differ quantitatively. Of the two the urea derivatives seem to be the least hemotoxic in man.

Recently HELLMANN's group at the Imperial Cancer Research Fund Laboratories in London claimed for a group of a series of bisdiketopiperazines or EDTA derivatives considerable anti-tumour activities (CREIGHTON et al., 1969).

A number of investigators have taken interest in the naturally occurring polyamine spermine, because it exerted some carcinostatic action in experimental cell cultures in the presence of spermine oxidase and in L 1210 mice. ISRAEL et al. (1966) prepared a number of spermine analogues and studied the chemistry and biological activity of the whole group. Their colleague ALARCON (1964, 1966, 1968) claimed that the truly active principle is acrolein, an unsaturated aldehyde, $CH_2 = CH.CHO$ (a break-down product of the dialdehyde formed from spermine by the action of oxidase). If valid activities for the spermine oxidase alone could be demonstrated, probably acting on endogenous polyamine, then a further enzyme preparation might join the present series.

Claims for anti-tumour activity of hemolytic streptococci have been made by a number of laboratories. The possibility is not yet completely excluded that this biological activity is due to the production by the microbe of streptolysin S. If this proves to be true, then this observation should be placed into the next subchapter.

2. Antibiotics, alkaloids and other natural products

After the discovery, the start of commercial production and the first applications of the early antibiotics, for instance penicillin and streptomycin, hope was expressed by a number of medical scientists that of these novel products of fungal and microbial origin one or the other might attack abnormal cells as effectively as bacterial cells. Now, twenty years later, a very large number of cultures and their culture fluids, as described in detail by STOCK (1966b) in his excellent survey of this field, have been tested (particularly by the Cancer Chemotherapy National Service Center, Bethesda, and some Japanese laboratories), and a number of materials were found to have pronounced anti-leukemic and anti-tumour activities. This material is dealt with also in an earlier review by WOODRUFF and MILLER (1963). A considerable array of more specialized reviews on individual antibiotics are at the disposal of the interested reader and student of this chapter of carcinochemotherapy, and one can obtain a clear picture of the relatively young history of the search, the success and failures, and the potentialities for the future. May it suffice to quote for actinomycin (Dactinomycin) BROCKMANN (1960), FARBER (1968) on the clinical and biological properties of this antibiotic and MODEST et al. (1963) on the place of polypeptidic antibiotics in the general group of peptides and proteins as inhibitors. One must mention also the contributions of the eminent microbiologist WAKSMAN (1963) to whom the cancer research workers, clinicians and patients owe so much. WAKSMAN and FURNESS (1960) edited a review on the use of the actinomycins in the treatment of tumours.

8*

Although these surveys are now somewhat outdated, they allow a proper assessment of the historical background of this significant advance in carcinotherapy. As indicated before, the countries which were in the forefront of anti-tumour antibiotics research and dedicated a great amount of effort to finding new types of anti-neoplastically active substances were the United States and Japan. This does not mean that in other places, such as Germany and Britain, only an inconsiderable amount of work was done, particularly on chemical problems and those connected with action mechanisms. MAUGER and WADE (1966) even synthesized a compound which represents a combination of the tricyclic actinomycin carrier with the antibiotic decapeptide gramicidine S. However, this first synthetic analogue of actinomycin showed little activity, indicating the necessity for the correct peptide structure. Meanwhile MEIENHOFER et al. (1966) have opened up an elegant synthesis of actinomycin D itself and MODEST et al. (1969) are studying the biochemistry of the antibiotic and the dependence of the anti-tumour action on certain features of the whole molecule. So far no preparation of an analogue with diminished toxicity but the same anti-tumour effects has been discovered. FARBER (1966) reported summarily on the effects of the drug of natural origin, and two years later a chapter in one of the volumes of Recent Results in Cancer Research, "Tumours in Children", edited by MARSDEN and STEWARD (1968), mentions again actinomycin, together with vincristine, as one of the chemotherapeutic approaches in pediatrics.

Mitomycin C, carcinophyllin and H-bleomycin (UMEZAWA, 1968) are three of the more outstanding Japanese antibiotics. Some are discussed in the reviews mentioned above. A newer, most interesting compound is mentioned by ONO et al. (1968) and MAEDA et al. (1966) under the name of neocarcinostatin, isolated from *Streptomyces carzinostaticus*. It is a protein and its mode of action seems to rest on specific DNA degradation. So far this growth regulation has been observed in microorganisms; its occurrence in higher animals awaits confirmation.

In going over a number of Cancer Chemotherapy Reports, especially those published between the years 1966 and 1968, one can pick up very useful records of effects of older and newer antibiotics against experimental or clinical neoplasms. Whether it concerns daunomycin (rubidomycin) (VENDITTI et al., 1966) in its behaviour towards experimental tumours, or its behaviour as a labelled compound in animals (RUSCONI et al., 1968), or lesser known antibiotics, such as streptonigrin from another *Streptomyces* and recently broadcast in a Chemotherapy Fact Sheet by the National Cancer Institute (CARTER, 1968), or porfiromycin which is a methyl derivative of mitomycin C, these articles demonstrate the interest still maintained in this group of agents, although none has the ideal properties for which so much hope was entertained some years ago.

Apart from the alkaloids of the periwinkle plant (*Vinca rosea*), especially vincristine, no novel, therapeutically startling higher plant product has emerged from any of the numerous laboratories that are preoccupied with natural materials. A whole symposium was held on the subject of the vinca alkaloids in Memphis, Tennessee (SULLIVAN, 1968), which covered everything from biochemistry to detailed clinical work of these chemically complex materials (see also Proc. 1st Symposium Europ. Cancer Chemotherap., June 1965). Their place in combination therapies of leukemias and lymphomas appears to be quite secure at present. A small number of plant products, such as the cucurbitacins, aristolochic acid and the older colchicine and podophyllotoxin are briefly described and annotated in a review by STOCK (1967a), called 'Other Tumour Agents'.

Climbing higher up the evolutionary ladder — or lower, depending on the viewpoint — it should be mentioned that clam extract, bacteria and animal materials, e.g. liver, have been explored for their elusive action against malignant tissues. It is often difficult to decide whether any of these materials are active due to some toxic action, or due to what will be discussed below as "homeostatic regulation". Indeed some of the observed effects may even involve the immunological responses of the host, or the stimulation of his lysosomes, or combinations of all of these built-in regulatory mechanisms and the host's defence systems (see WOLSTENHOLME and KNIGHT, 1969).

Concerning the extract from *Merceneria merceneria*, the common clam, observations have been reported during the Toyko Congress (LI et al., 1966). The use of bacteria, such as hemolytic *Streptococci* or *Clostridia* takes one into the so-called biological type of tumour therapy, although in the former case (OKAMOTO et al., 1966) the effects may be produced by streptolysin S. The interest in anaerobic microbes, e.g. *Clostridium butyricum*, as investigated by GERICKE et al. (1967b), rests on the metabolic activity of the organism in experimental tumour material *in vitro*, where lysis takes place. Unfortunately, human material did not respond in the same manner. However, it is not impossible that propionic acid bacteria may be more effective under *in vivo* conditions. Recent work by the same author (1969) extends to mycoplasms.

As mentioned above, there are several investigators who have tested extracts of animal tissues. To give two examples, one from Japan and one from the United States: SUGIHARA et al. (1966) studied a component of bovine liver, and BARDOS et al. (1966, 1968) screened a great number of various tissue fractions against bacteria and three experimental tumours. But one can take it for granted that anyone engaged in cancer research has in his team at some time someone who has experiments going along this line, whether it is in Germany, Sweden, or in other member countries of the U.I.C.C.

IV. Chemo-biological therapies

A. Suppression

1. Immunological approaches

It is certain that some cancer research workers and immunologists will raise strong protests against the inclusion of a subchapter on cancer immunology in a review of the field of carcinochemotherapy. Taking a very traditional standpoint, they may be right (perhaps Paul Ehrlich would be an exception, if he were still alive today). Molecular biology is changing rapidly into genetic and cellfunctional biochemistry and biophysics (see BERGEL, 1966a) concerned with nuclear events and their relation to cytoplasmic particles, cell membranes, cell division and cell to cell interactions through macromolecules. It would be rigid on the part of the orthodox and irresponsible on the part of medical scientists who aim at an understanding of the totality of health and disease, if attention to low-molecular carcinochemotherapeutic agents were separated from the attention given to macromolecules of biosynthetic origin (cf. some of the enzymes mentioned in this review). Who knows how soon antigens and antibodies, immuno-stimulators and messengers, and nucleic acids that seemingly govern the regulation of biosynthetic processes of circulating and cell-bound host defences (see ALEXANDER, 1968), will be open to synthetic attacks by organic chemists? Although the whole subject of carcinoimmuno-therapy is still at a very early stage, it is not because of an overestimation of the molecular and an underestimation of the biological events that the reviewer includes this topic in the present essay. It is true that, with few exceptions, those who are actively helping in a revival of cancer immunology are medically or biologically inclined. Nevertheless, there are many physico-chemical and chemical problems interwoven with the welter of biological phenomena, so that teams consisting of scientists belonging to different disciplines have been responsible for the most outstanding advances during the last few years.

If one could answer the questions why cancer immunology had such a set-back during the first half of this century and why there was this renaissance during the last decade or so, the answer would assist in illuminating more sharply its present problems, its limitation and its prospects. To such an end, it is helpful to consult the publications of at least some of the scientists active in this field. There must be many laboratories in many countries that have published results which at some future date may prove to be the true turning point towards a resolution of the intricate difficulties, and which are not quoted in these pages (see NOSSAL, 1969). For sheer practical reasons this brief review within a review will be based on the utterances of the Chester Beatty group (HADDOW, 1965; ALEXANDER and HAMILTON FAIRLEY, 1967; DELORME and ALEXANDER, 1964, etc.), Mathé's Institute in Villejuif (MATHÉ

et al., 1965), SOUTHAM's team in New York (1965) and DAY's in Durham,
N. C. (1965), KLEIN's group in Stockholm (1960) and GERICKE's Krebsforschungslabor in Frankfurt (1968). The contributions of others (and the
list is far from complete), such as the Russian (ZILBER), French (GRABAR),
British and Australian teams (BALDWIN, BURNETT, GORER, MEDAWAR,
MILLER, NOSSAL, WOODRUFF) can be culled from bibliographies attached to
papers, the references to which will be cited below. One point should be
stressed, namely that nearly all of the work up to now is of an experimental
nature. Even that extending to man (see MATHÉ et al., 1965; ALEXANDER and
HAMILTON FAIREY, 1967; SOUTHAM, 1968) represents clinical research and at
this moment cannot be compared with the more traditional run of carcinochemotherapy. In fact, some of the work done on patients has produced some
objections by some members of the medical world (see SCANLON et al., 1965).
ALEXANDER (1968) finishes a postscript to a review on "Immunotherapy of
Cancer": "At best, immunotherapy may become an adjuvant for the removal
of residual tumour cells after other forms of treatment have been used"
(see MIHICH, 1969).

The multi-etiology of cancer or, more correctly, cancers, concomitant
with great variations in the degree of immunological competence of the hosts,
makes the present state of knowledge and research approaches a subject
appropriate for the expert only. The existence of humoral or circulating or
free and, on the other hand, cell-bound immunobodies may confuse the
interested outsider, and so does the complex and inconsistent nomenclature.
With some exceptions, a number of workers in the field investigate and discuss
one to the exclusion of the other, without distinguishing properly for the
non-expert between immune lymphocytes and antibody-containing fluids and
tissues. But this confusion should be reduced in the course of the next few
years, for instance by an increasing habit of applying various techniques to
numerous samples of experimental and human neoplasms. As MATHÉ (1969b)
suggested, advanced courses on methods for the detection of specific tumour
antigens should be held to teach tests, e.g., for cytotoxicity, complement
fixation, immune adherence, immuno fluorescence, immuno-diffusion, and
others. In a brief review GERICKE (1968) mentions a number of serological
assays, among them passive haemagglutination, and immunoelectrophoresis
and the opsonization test.

The state of basic knowledge up to 1961 is well represented by three papers
read at a symposium on Biological Approaches to Cancer Chemotherapy at
Louvain under the auspices of UNESCO and WHO. There, G. KLEIN (1961b),
GORER (1961) and ZILBER (1961) delineated the elementary design of what
laboratories during the subsequent eight years pursued in their researches,
expanding the subject in many directions. Several general principles emerged,
namely that, in spite of the close relation between normal and cancer cells,

there exist antigenic dissimilarities between the two, particularly pronounced in chemically induced tumours; that genetic properties play an important part in experimental tumours; and that a sharp distinction must be drawn between host resistance directed against an autochthonous tumour and implanted ones (see KOLDOVSKY, 1969). Further descriptions of background and practical conclusions are given with many details in DAY's Immunochemistry of Cancer (1965). He comes to these conclusions: There is a choice of hetero-, iso- and autoimmune types of anti-sera which require thorough purification before their use as therapeutic agents. Preventive immunity to virus-induced cancers is basically more rational than a similar arrangement for non-viral cancers. The overall impression of these summarizing remarks is, of course, the preoccupation of the monograph with circulating immunobodies.

In contrast, ALEXANDER and his colleagues (1968a and b) and DELORME of the same group (1967), following KLEIN's (1961b and 1966) predilection for cellbound antibodies, concentrated on problems other than immune sera. But prior to giving their results a brief airing, it might be interesting to look at the writings of SOUTHAM (1965), because he applied in autochthonous cancers in man the techniques of autotransplantation, homotransplantation, skin tests and serology to answer the question whether there exist promising host defences. Naturally, the biological complexity, as he says, of cancer and the limitations of clinical research, as compared with animal experimentation, make it difficult to arrive at unequivocal conclusions. Nevertheless, there are signs that there exist host defences which may act upon cancer-specific antigens, probably through cellular mechanisms. How complex a situation one has to deal with is shown by his description (SOUTHAM, 1968) of the case of a cancer patient who is immune against homografts but still succumbs to a tumour established in the body. Like ALEXANDER (1968a and b), SOUTHAM comes to the conclusion that there are, in spite of this, defences against autochthonous cancers but that such defence forces can deal only with small numbers of residual malignant cells remaining after other therapies, including surgery. Interesting results are discussed by ALEXANDER and HAMILTON FAIRLEY (1967) in connection with the relation between cellular resistance and delayed hypersensitivity. It appears that, in malignant diseases in man, e.g. chronic lymphocytic and other leukemias, Hodgkin's disease, etc., the delayed hypersensitivity response to foreign antigens is impaired. A similar impairment can be observed also in cases of carcinomas and sarcomas. This may mean that, in man, cancer immunity systems[3] are mainly of a cellular nature, very likely related to the degree of activity of competent lymphocytes, and that they are deficient in cancer patients (NOSSAL, 1969).

---

3 In Recent Results in Cancer Research (Springer-Verlag) MATHÉ will write about 'L'Immunothérapie des Cancers".

Leukemias and Burkitt's lymphoma are very interesting subjects for studying immunity phenomena in cancer. A whole session during a Conference on Acute Leukemia and Burkitt's Tumour in Rye, New York, was devoted to the immunological aspects connected with these diseases. From the KLEINS (1967) via a galaxy of immunologists to MATHÉ (1967) and SOUTHAM (1967), every shade of opinion was represented. In the summary (SOUTHAM, 1967), the property of Burkitt's tumour of having a tumour-specific surface antigen was underlined. ALEXANDER (1967) attempted clarification of terminology: it was pointed out that there was a difference between active immunization = administration of an antigen; passive immunization = administration of an antiserum; and adoptive immunization, possibly meaning that it results from the administration of "immunocytes". ALEXANDER'S report on successful treatment with "messenger cells", producing cytotoxic lymphocytes, was considered to be a promising approach, and MATHÉ'S discussion of the role of immunology in the treatment of leukemias and hematosarcomas (1967) touched on the administration of leukemic cells, a form of active immuno-therapy, and (in animals) on the administration of immunocompetent cells as a form of an "adoptive" therapy.

ALEXANDER (1968a) proposes to restrict the term immunotherapy to procedures which affect the growth of tumours where there are no histo-incompatibilities; experiments with tumours, transplanted in allogeneic tests, provide information about homograft activity and *not* tumour immunology. The following approaches are under active consideration:

(1) Active immunization with cells prepared from the tumour to be treated; this means possibly the host's own immune reaction is enhanced.

(2) Nonspecific stimulation of the host's reticuloendothelial system; this applies mainly to grafted tumours.

(3) Specifically immunized lymphocytes are lethal for tumour cells by immediate contact; this may involve a process of passive cellular immunity with cytotoxic lymphocytes and messenger cells which could confer on the host the capacity to produce an increased (improved?) immune response.

(4) Ribonucleic acid, extracted from lymphocytes immunized against the tumour to be treated, may be the carrier of the message mentioned under (3) above; sheep and goats have been used for these procedures.

(5) Immune sera: these are the subject of the investigations by GORER (1961) and ZILBER (1961) and are mentioned in the monograph by DAY (1965).

2. Anti-viral approaches

The question cannot be answered yet whether, after the virus etiology of certain human cancers, particularly leukemias and lymphomas, has been established beyond doubt, immunological means will be used to remove the viruses in a way similar to their eradication in other viral diseases (see STANLEY, 1969).

Returning once more to the Rye Leukemia Conference of 1967, another session was dedicated to the provision of evidence for viruses in the two named morbid states. Whether herpes-, reo- or simian-viruses are indictable cannot be pronounced with certainty. But the development of immunological tools in virology generally and any advance that could lead towards a virus chemotherapy might help in developing this rather poorly cultivated field. So far the most effective experimental group of compounds has been found among a number of antagonists of pyrimidines, purine and folic acid metabolism (TIMMIS and WILLIAMS, 1967, p. 176—194). Should these and related compounds prove ineffective, one would have to wait for pronounced success achieved by using drugs such as benzimidazoles, thiosemicarbazones, indole derivatives (Marboran), triazines, steroids or macromolecules (BERGEL, 1965; SADLER, 1962). Otherwise progress with natural or even semi-synthetic interferons might offer therapeutic techniques so far not put to extensive tests (BURKE, 1962).

3. Biological approaches

The carcinolytic effect of bacteria has been mentioned on previous pages. The same applies to randomly prepared extracts from animal tissues. WHISSON (1967) has recently contributed some ideas to the possible utilization of embryonic tissues producing in tumours a kind of redifferentiation process (see below). This approach is reminiscent of the investigations of a number of workers who will be mentioned without references, since these can be retrieved in the volume of the Ciba Foundation Symposium on "Cell Differentiation" (DE REUCK and KNIGHT, 1967). They are SEILERN-ASPANG and KRATOCHWIL, and the WOLFFS and AMBROSE et al. Other experiments are progressing that make use of sperms, viruses and similar carriers of biomacromolecules. Whether these efforts will lead to tangible results is an open question. There are indications in the literature (apart from the report on the use of specific RNA as a messenger substance for host lymphocytes) (ALEXANDER, 1968 b) that polyanions do show sometimes antimitotic activities (REGELSON, 1968 and Rosso et al., 1969). To summarize: from a pragmatic point of view, this mixed chapter on "suppression" describes researches of a highly active nature. But "active" refers more to the fervour of the participants than to practical applications. Only the future will tell whether some of these efforts will be fruitful.

B. Balance and regulation
1. General remarks

All the modes of chemical and chemo-biological cancer treatments so far discussed relied on the cytotoxic action of chemicals, biochemicals, antibodies or competent cytotoxic or messenger lymphocytes (see ALEXANDER, 1967).

Supplementing the very brief remark made in connection with biological approaches, it is now desirable to turn to the possibility, vigorously denigrated by several leading cancer research workers, of reversing the process of carcinogenesis and returning the cells, carrying all the hallmarks of malignancy, to a more mature state. Normally, mitosis, DNA synthesis, loss of differentiation and polyploidy are under the strict control of a system or systems that, in parallel to the fine balance of the autonomic nervous system (functional), could be called homeostasis regulating systems. Do these exist solely in the form of interlocked inhibitory (feed-back) and stimulatory processes, or are there any entities hidden in the cells and tissues which could be characterized chemically and biochemically?

G. KLEIN, one of the leading cancer immunologists, remarked once in a discussion after P. GORER's paper on isoantigens (1961), read during a symposium in Louvain: "Two strains of mouse cells studied by SANFORD et al. (1948) have been derived from the same single (so called normal) cell. They produce a high or low incidence of cancer, when implanted back into C3H mice" (see also DE SOMER and PRINZIE who give 97% for high and 1% for low incidence, 1961). If one can accept this as a phenomenon that is generally observable and not only caused by immunological reactions, and if this represents something more akin to an "Avery type of bacterial transformation" than to changes of normal mammalian cells into malignant cells by polyoma virus (see Fig. 1; STOKER, 1964; SACHS, 1966), then, and only then, could this serve as evidence for a biological move in opposite directions, one toward carcinogenesis, the other toward normality or, at least, preserved maturity. Whether this happens every millisecond, hourly, daily or at any stage of the cycle of dividing cells, was among other problems the subject of a discussion at a joint Ciba Foundation-Wellcome Trust Conference on "Homeostatic Regulators" (WOLSTENHOME and KNIGHT, 1969). On this occasion the question of hormones and chalones as contributors to homeostatic balance was touched upon, and the control of carcinogenic processes and, hence, their possible arrest and reversal formed only a small part of the total proceedings. For this reason and because hormones have been investigated for many years in cancer research and outside and chalones have come into focus recently, both will be treated briefly below.

The difficulty all-round is a proper definition of homeostatic regulators and homeostasis, particularly if one wishes to apply these terms to cellular and tissue events. Physiologists like to retain the original meaning given to it by CANNON (1929, 1932) who equates homeostasis with stability of the internal environment (BEST and TAYLOR, 1945), especially that of the autonomic effector cells and their functional equilibrium. If one wishes to extend this definition, it ought to mean maintenance of order, not only of function but also of cellular and growth patterns and of chemical compositions of fluid and tissues, in face of disorder, perturbation of the environment (including that which obtains inside the body vis-à-vis its organs, and cellular units and components), in form of injury, starvation, sudden changes in temperature, microorganisms, viruses, mutagenic and carcinogenic agents. Narrowing the borderlines, inside which such adjustments are essential for the survival of any part of the organism, one has to include wear and tear and repair of tissues, growth arrest, cell division (mitosis), cell death and maintenance of all cell characteristics, threatened by external and internal assaults. Stedman's Dictionary (1966) adds to the term homeostasis the processes themselves through which such cytological and body equilibria are preserved inside maximal and minimal limits. While at the level of physico-chemical and biochemical events, equilibria are maintained automatically by electron exchanges and feed-back and feed-forward mechanisms (Iversen, 1969); subcellular, cellular and tissue events depend for their normal and healthy states (which of course alter in character during the life span and reproductive activity of the organism) on the order inside the cell, on the rate of mitosis, the ratio of proliferation to cell death, and mitotic arrest. To add only two more examples, further regulation arises also from normal immunological and, as mentioned already, hormonal activities and balances.

2. Hormones

These endocrine stimulators (Clayson, 1962) function only in certain target tissues controlled mostly through the activities of the hormones of the pituitary. There are the sex-organ effects of the estrogenic, androgenic and progesteronic agents; the adrenal medullary hormones (adrenaline, nor-adrenaline, etc., as part of the autonomic system and more recently studied in the brain and pituitary) and the corticoids. Most of these are linked closely together by a built-in homeostatic arrangement; for instance, thyroid-stimulating hormone (TSH) versus thyroxine, adrenocorticotrophic hormone (ACTH) versus cortical adrenal hormones and gonadotrophic hormones (LH luteinizing and FSH follicle-stimulating) versus the sex hormones. Their imbalance can cause marked alterations in cells and tissues of hormone-dependent parts of the body, which may lead to pathological changes or to therapeutic effects. There is no need at this point to go into details of action mechanisms. Suffice it to remind the reader of a review by Wade (1967) of hormones used for carcinochemotherapy. If they are withdrawn by surgical means or suppressed, this amounts to a "negative chemotherapy," as Boy-land (1961) suggested. He and Clayson quite rightly have emphasized the regulation of growth processes; they mention analogues of testosterone, estrone and corticosteroids, some with fluoro-substituents, and the leukemia-standby ACTH. The specificity of these compounds for certain tissues allows for a kind of selectivity in their application in carcinochemotherapy.

During the last two or three years, cancers, such as that of the breast and its metastases, were treated with Δ'-testololactone (PAPAIANNOU and VOLK, 1966); 6α-chloro-16α-methylpregn-4-ene-3,20-dione (GOLDENBERG et al., 1966) and estradiol undecylate (KENNEDY, 1967). Other compounds for studies in patients with breast cancer included an androgen, 7α-methyl-19-nortestosterone (O'BRYAN and TALLEY, 1966), while in children with acute leukemia a progesterone, fluorometholone, was found as effective as prednisone. The latter is often used in combination therapy (HARTMANN et al., 1966). A relatively good response was achieved with 6α-methylpregn-4-ene-3,11,20-trione (JOHNSON et al., 1966).

It is not unlikely that other hormones, whether steroidal or peptidic, will be synthesized during the next few years and may show improved effects and reduced toxicities. However, there is no immediate prospect that cancers can be cured in all cases of hormonally dependent tissues, whether by hormone withdrawal or shift in the disturbed homeostasis. Perhaps one day hormones in combination with other regulators, to be mentioned presently, and combined with surgical procedures may find a revival as remedies of malignant disease.

3. Chalones and other potential regulators

As mentioned earlier, regulators which might act homeostatically in tissues, independent of hormonal control, with respect to cell division and growth and against all the potential abnormalities arising from an assault by a number of adverse factors, could exist in the form of entities or of complex systems. A few years ago. BULLOUGH (1965) and IVERSEN (see 1969) developed, independently and jointly, the idea that growth inhibitory factors were present in certain tissues. They were called "chalones" (from the Greek, meaning to 'slacken'). The first organ studied was the skin (see BULLOUGH and LAURENCE, 1968a; IVERSEN, 1969). IVERSEN is rather cautious in his interpretation of the results so far available (pilot manufacture from pigskin was carried out under the supervision of HONDIUS BOLDINGH and LAURENCE, 1968, Messrs. Organon, Holland). On the other hand, BULLOUGH and LAURENCE (1968a), emphasizing the so far inexplicable requirement for adrenaline and a glucocorticoid hormone if the goal is full inhibitory effect on epidermal mitosis, claim (1968b) that tumour cells have a reduced intracellular concentration of the chalone. Further work by the London Mitosis Research Laboratory (BULLOUGH and LAURENCE, 1968c) led to the discovery of a melanocyte chalone of which the authors maintain that it inhibits specifically mitosis in melanocytes *in vivo* and *in vitro*. Support for this has come from U. MOHR et al. (1968) in Germany who produced experimental proof for an effect of the active principle, which is tissue-specific but not species-specific, on Harding-Passey and hamster (amelanotic) melanomata. The chalone is different from the epidermal one, also described by IVERSEN (1969). All authors agree that the active substances are unstable. Finally, two Finnish workers (Rytömaa and Kiviniemi, 1968a, 1968b) have published results which indicate the existence of a granulocytic chalone. This again is tissue-specific and inhibits DNA synthesis both in normal and leukemic cells but RNA synthesis only in normal granulocytic progenitor cells. A lung chalone has been recently discovered by Simnet et al. (1969). One should obtain a better understanding of the whole situation when further work discloses the chemical character of the chalones, their dose-response relationship, and whether they act as true homeostatic regulators or otherwise, under normal and abnormal, benign or malignant conditions. Compared with growth factors, such as the nerve growth factor (NGF) and the epidermal growth factor (EGF) of which the former produces its effects in very low concentrations (10^{-14}g·ml^{-1} *in vitro*; Banks et al., 1968), the chalones in their present form of impurity appear to exert their activity on mitosis only when given in relatively large concentration ranges (about 10^{-3}g·ml^{-1}).

Taking an overall view, one must not forget that other control systems, such as those of the cell nucleus (repressors, de-repressors, enzymes, histones), of lysosomes acting as cellular debris removers (de Duve, 1968; Jaques, 1969), of mitochondria (see Nass, 1969), and perhaps in the form of isoenzymes, immunobodies (see previous subchapter), of surface and membrane forces and materials, and of cell to cell interactions may play a more or less vital part in the homeostatic apparatus of cells and tissues. Whether interferons (Burke, 1962), the Wolff differentiation-preserving factors (Wolff and Wolff, 1967), the thromboplastic materials of human tumours (Boggust and O'Meara, 1966) have a niche in what one might justifiably assume to be a multimembered arrangement, cannot be decided yet. But together with hormones, and leaving the reversibility of malignant to normal states at present outside the consideration of this review which the reader may judge to be too speculative anyhow, one or the other of these agents may be the long-sought-for control factor(s) missing or reduced in neoplastic cells!

V. Summary

1. Chemotherapy from the morbidity point of view

Clinicians, up to a relatively recent date, tried indiscriminately everything, which the laboratory produced, as cytostatic or carcinolytic compounds: that is, those who had any inclinations to do something about inoperable or radiation resistant cases. Leukemias formed an exception, because Farber, in 1947, had started the first successful carcinochemotherapy on a rational and sound basis. When, after this first attack on acute forms, chronic myelogenous (or granulocytic) leukemia became controllable by orderly administration of busulfan (Myleran), the further development of therapy in this group of diseases passed through the stages of application of a variety of dose schedules and combination therapy (ZUBROD, 1967; BURKE et al., 1968; TUCKER et al., 1968) to the application of total and systematic patient care or supportive therapy (platelet replacement, treatment of infection, etc.) (DJERASSI and FARBER, 1965; FREI, 1965, 1969; KLEIN et al., 1965; FARBER, 1967).

As hinted at above, the situation with solid tumours is different. True enough, choriocarcinoma (HERTZ et al., 1961) responds well to antimetabolites, such as methotrexate, sometimes in combination with 6-mercaptopurine or actinomycin (Ross et al., 1962); multiple myeloma therapy showed some success with melphalan and cyclophosphamide; Burkitt's lymphoma is percentage- and time-wise the most susceptible neoplasm to a number of alkylating agents, although not infrequently recurrence produces disappointments; Wilm's tumour has been treated by FARBER et al. (1966, 1968) with Dactinomycin with excellent results in some cases. But all the other forms of tumours, except breast cancer with its hormonal therapy and adrenalectomy, were subjected to a rather haphazard chemotherapy, mostly away from cancer research centers. As mentioned before, anything that became available was applied, often on no other basis than that of inhibition of experimental transplanted tumours in rodents. One hopes fervently that this phase is over, as it brought carcinochemotherapy into discredit and did not help the essential close relationship between clinician and the laboratory scientists. Like the treatment of leukemias, solid tumours merit a more rational approach, maybe on the basis of biochemical behaviour of the tumour without and after the impact of the drug. In the opinion of the reviewer, it is necessary to concentrate on a selected number of drugs or combinations of such drugs, or of newer treatments as described in Table 1, p. 93. Not that one may not pay careful attention to the practical developments of adjuvant therapies in the form of host defences, or active immunization, or perhaps of homeostatic regulators, such as hormones or chalones. It is expected that monographs like those on Tumours in Children (MARSDEN and STEWARD eds., 1968); the treatment of Hodgkin's disease (see HANCOCK, 1968) with a whole treatise on chemotherapeutic substances (ANGLESIO, 1969); the Treatment of Skin Cancer (FREEMAN und KNOX, 1967) with its careful analysis of surgical, radio- and chemo-therapies will be followed in the near and distant future by others, with material allowing an even more critical assessment of the type of carcinochemotherapy to be applied, showing the best possible prospects.

2. Future possibilities

It is easier to be a historian than a prophet, so a former President of Harvard said. Future symposia on either the national or international level will, in their programs, include attempts to look round various corners. The Conference in New Jersey last spring organised by GARATTINI et al. (1969;

see also MANDEL, 1969) included immunological facts and guesses, glanced at the regulatory mechanisms and provided fresh fodder for the lab-stables, running alkylating agents and antimetabolites. Like the course held under the auspices of WHO and the International Agency for Research on Cancer (MÜHLBOCK, 1969) on improvements in the use of experimental animals, so newer advances in the knowledge of the effects of drug schedules and combinations (see NATHANSON, 1966) will play an increasing part in work and reporting. ZUBROD (1967) pointed out that it was not the addition of new agents that produced a rise in the mean survival time of patients with acute lymphocytic leukemia but a change of the clinical planning and facilities. In the pressing timetable of cancer research, it is a long time ago that ENDICOTT (1965) laid down a program for his Institute and its collaborators all over the United States (see also ZUBROD et al., 1966). The virus field, except for progress on the experimental side, is at this moment at a practical standstill with regard to the human patient. This reviewer remains a loyal believer in continued efforts in the field of applied biochemistry and biology (see STOCK, 1967a and b; SAHASRABUDHE, 1967) mainly of human cancers (see also HARRAP and JACKSON, 1969; DAVIS, LOMBART et al., 1969). He is possibly lacking caution, if he foresees a continuation of investigations into the psycho-physiological aspects of cancer. If the basis for such aspects exists, and if some cancers can be counted among psychosomatic diseases, then the logical outcome of this existence should be systematic and painstaking research into psychosomatic treatments (see BAHNSON, 1969).

References

ADAMSON, R. H., FABRO, S.: Some studies with asparaginase, asparagine and asparagine analogues. Proc. Amer. Ass. Cancer Res. 9, 2 (3) (1968).
— — Antitumor activity and other biological property of L-asparaginase. A review. Cancer Chemother. Rep. 52, 617—626 (1968).
ALARCON, R. A.: Isolation of acrolein from mixtures of spermine with calf serum and its effects on mammalian cells. Arch. Biochem. 106, 240—242 (1964).
— Acrolein II. Studies on the fluorescence produced by acrolein, spermine and related compounds with resorcinol and calf serum. Arch. Biochem. 113, 281—287 (1966).
— Fluorometric determination of acrolein and related compounds with m-aminophenol. Analyt. Chem. 40, 1704—1708 (1968).
ALEXANDER, P.: Immunotherapy of leukemia. The use of different classes of immune lymphocytes. Cancer Res. 27, 2521—2526 (1967).
— Immunotherapy of cancer. Experiments with primary tumours and syngeneic tumour grafts. In: Progr. Exp. Tumor Res. (Basel) 10, 22—71 (1968a).
— Treatment of experimental tumours with immune lymphocytes. In: Proc. Int. Symp. on Gammapathies, Infections, Cancer and Immunity, p. 81—83 (V. CHINI, S. BONOMO and C. SIRTORI, eds.), 1968b.
— HAMILTON FAIRLEY, G.: Cellular resistance to tumours. Brit. med. Bull. 23, 86—92 (1967).
ANGLESIO, E.: The treatment of Hodgkin's disease. In: Recent results in cancer research, vol. 18, p. 30—42. Berlin-Heidelberg-New York: Springer 1969.

BAHNSON, C. G.: Chm., Second Conference on Psychophysiological Aspects of Cancer May 1968. Ann. N.Y. Acad. Sci. in press (1969).

BANKS, B. E. C., BANTHORPE, D. V., BERRY, D. V., DAVIES, H. ff. S., DOONAN, S., LAMONT, D. M., SHIPOLINI, R., VERNON, C. A.: The preparation of nerve growth factors from snake venoms. Biochem. J. **108**, 157—158 (1968).

BARDOS, T. J., GORDON, H. L., CHMIELEWICZ, Z. F., KUTZ, R. L., AMBRUS, J. L.: Screening of animal tissues for inhibitory substances. Abstr. Papers, SO 596, IXth Int. Cancer Congr., Tokyo 1966.

— — — — — Systemic investigations of the presence of growth inhibitory substances in animal tissues. Cancer Res. **28**, 1620—1630 (1968).

BARRY, V. C., CONALTY, M. L., O'CALLAGHAN, C. N., TWOMEY, D.: Anticancer agents III. Synthesis and anticancer activity of some bisthiosemicarbazones and thiosemicarbazones. Proc. roy. Irish Acad. B **65**, 309—324 (1967).

— McCORMICK, J. E., McELHINNEY, R. S., McINERNEY, M. R., O'SULLIVAN, J. F.: Anticancer agents. I. Structure-activity relationships in a series of oxypolysaccharide-thiosemicarbazide derivatives. Proc. roy. Irish Acad. B **64**, 335—354 (1966).

— — — — — TWOMEY, D.: The relationship of chemical structure to anti-cancer activity in substituted thiosemicarbazide derivatives of dicarbonyl compounds. Abstr. Papers SO 550, IXth Int. Cancer Congr., Tokyo 1966.

— — O'SULLIVAN, J. F.: Anticancer activity of bisthiosemicarbazones of methylglyoxal. Cancer Res. **26**, 2165—2168 (1966).

BERGEL, F.: Chemistry of enzymes in cancer. Springfield (Ill.): Ch. C. Thomas 1961.

— Development, possibilities and prospects of alkylating agents in chemotherapy of cancer (ed. P. A. PLATTNER), p. 21—31. Amsterdam: Elsevier Publ. Co. 1964a.

— Optical stereospecificity of anti-cancer agents. Farmaco **19**, 99—109 (1964b).

— Whence and whither chemotherapy. Proc. roy. Instn. **40**, 355—374 (1965).

— Zusammenfassung und Ausblick. In: Molekulare Biologie des Malignen Wachstums. Berlin-Heidelberg-New York: Springer 1966a.

— In place of experimental animals. New Scientist, 9th June, 646—647 (1966b).

— Current advances in cancer research. Nursing Times, London, December 1967.

— HARRAP, K. R.: Future possibilities for the development of treatments in leukemia. Cancer Res. **25**, 1643—1648 (1965).

— — SCOTT, A. M.: Interaction between carbonyl groups and biologically essential substituents, part IV. An enzyme model system for cysteine desulphydrase. J. chem. Soc. 1101—1112 (1962).

— JOHNSON, J. M., WADE, R.: Antitumour activity of a series of acetyldipeptide esters containing melphalan, p. 241—245. In: Proc. 6th Europ. Symp., Athens, Pergamon Press., London 1965.

BEST, C. H., TAYLOR, N. B.: The physiological basis of medical practice. 4th ed. London: Baillière, Tindall & Cox 1945.

BOESEN, F., DAVIS, W.: Cytotoxic drugs in the treatment of cancer. London: Edward Arnold 1969.

BOYLAND, F.: Hormone withdrawal in the treatment of cancer. Cancer Chemother. Rep. **13**, 7—8 (1961).

BRATZEL, R. P., ROSS, R. B., GOODRIDGE, T. H., HUNTRESS, W. T., FLATHER, M. T., JOHNSON, D. E.: A survey of alkylating agents. Cancer Chemother. Rep. **26**, 1—506 (1963).

British Association for cancer research, Autumn Mtg., Symposium on Aetiological Factors in Lymphomas, Sept. 1968.

— — — — — The Rationale and Assessment of Cancer Chemotherapy, Owen's Park, Manchester, April 1969.

BROCKMANN, H.: Die Actinomycine. Fortschr. Chem. org. Naturstoffe **18**, 1—54 (1960).

BROCKMAN, R. W.: Resistance to purine antagonists in experimental leukemia systems. Cancer Res. **25**, 1596—1607 (1965).

Brookes, P.: Reaction of alkylating agents with nucleic acids. In: Chemotherapy of cancer (P. A. Plattner, ed.), p. 32—43. Amsterdam: Elsevier Publ. Co. 1964.

Broome, J. D.: Evidence that the L-asparaginase activity of guinea pig serum is responsible for its antilymphoma effects. Nature (London) **191**, 1114 (1961).

— Evidence that L-asparaginase of guinea pig serum is responsible for its antilymphoma effect. J. exp. Med. **118**, 99—120, 121—148 (1963).

— Antilymphoma activity of L-asparaginase *in vivo*; clearance rates of enzyme preparations from guinea pig serum and yeast in relation to their effect on tumor growth. J. nat. Cancer Inst. **35**, 967—974 (1965).

— Studies on the tumor inhibitory action of L-asparaginase. Abstr. Papers, SO 584, IXth Int. Cancer Congr., Tokyo 1966.

— Mechanism of action of L-asparaginase. Proc. Conf. on L-Asparaginase, New Brunswick, N. J., 14th March 1968a.

— L-asparaginase: The evolution of a new tumor inhibitory agent. N. Y. Acad, Scie., Ser. II, **30**, 690—704 (1968b).

— Private Communication 1969. See also L-asparaginase, presented in: Z. B. Papanastassiou, Memorial Symposium on Current Trends in Cancer Chemotherapy. Med. Chem. Group, N. E. Section, Am. Chem. Soc. May 1969.

Bullough, W. S.: Mitotic and functional homeostasis. Cancer Res. **25**, 1683—1727 (1965).

— Laurence, E. B.: Epidermal chalone and mitotic control in the Vx 2 epidermal tumour. Nature (Lond.) **220**, 134—135 (1968a).

— — Control of mitosis in rabbit Vx 2 epidermal tumour by means of the epidermal chalone. Europ. J. Cancer. **4**, 587—594 (1968b).

— — Melanocyte chalone and mitotic control in melanomata. Nature (Lond.) **220**, 137—138 (1968c).

— — Control of mitosis in mouse and hamster melanomata by means of melanocyte chalone. Europ. J. Cancer **4**, 607—615 (1968d).

Burke, D. C.: Interferon — Some aspects of production and action. In: Symposium on Drugs, Parasites and Hosts (ed. L. G. Goodwin and R. H. Nimmo-Smith), p. 294—319. London: J. & A. Churchill 1962.

Burke, P. F., Lenhard, R. E., Jr., Owens, A. H. Jr.: Therapy for acute leukemia in adults with cytosine arabinoside, vincristine and prednisone. Cancer Chemother. Rep. **52**, 305—314 (1968).

Calvert, N., Connors, T. A., Ross, W. C. J.: Aryl-2-halogenalkylamines XXV. Derivatives of sulphonamide designed for selective deposition in neoplastic tissue. J. Europ. J. Cancer **4**, 627—636 (1968).

Cannon, W. B.: Organization for physiological homeostasis. Physiol. Rev. **9**, 399—431 (1929).

— Wisdom of the body. New York: W. W. Norton 1932.

Carter, S. K.: Streptonigrin. Chemother. Sheet. Nat. Cancer Inst., Nov. 1968.

Clayson, D. B.: Chemical carcinogenesis. Hormones and related compounds, p. 315—371. Hormonal therapy, p. 407. London: J. & A. Churchill, Ltd. 1962.

Cobb, L. M., Connors, T. A., Elson, L. A., Khan, A. H.: 2,4-Dinitro-5-Ethyleneiminobenzamide (CB 1954): A potent and selective inhibitor of the growth of the Walker carcinoma 256. Biochem. Pharmacol. **18**, 1519—1527 (1969).

Connors, T. A., Elson, L. A., Haddow, H., Ross, W. C. J.: The pharmacology and tumour growth inhibitory activity of 1-aminocyclopentane-1-carboxylic acid. Biochem. Pharmacol. **5**, 108—129 (1960).

— Whisson, M. E.: Cure of mice bearing advanced plasma cell tumours with aniline mustard. Nature (Lond.) **210**, 866—867 (1966).

Creighton, A. M., Hellmann, K., Whitecross, S.: Antitumour activity in a series of *bis*Diketopiperazines. Nature (Lond.) **222**, 384—385 (1969).

DAVIS, W.: Secr., A. Llombart, Hon. Chm. and the Co-ordinating Committee for Human Tumour Investigations. 3rd Int. Symp. on the Biol. Charact. of Hum. Tumrs., Madrid, April 1969.

DAY, E. D.: The Immunochemistry of cancer. Amer. Lect. Ser. Springfield (Ill.): Ch. C. Thomas 1965.

DE DUVE, C.: Lysosomes as targets for drugs. In: the Symposium on the Interaction of Drugs and Subcellular Components in Animal Cells (ed. D. C. CAMPBELL), p. 155—169. London: J. & A. Churchill, Ltd. 1968.

DELORME, E. J., ALEXANDER, P.: Treatment of primary fibrosarcoma in the rat with immune lymphocytes. Lancet 1964 II, 117—120.

— CONNEL, D. I., MIKULSKA, Z. B., ALEXANDER, P.: Immunological procedures for treating primary chemically induced fibrosarcoma in the rat and a murine-leukemia. In: Symposium on Specific Tumour Antigens (ed. R. J. C. HARRIS), p. 108—203. U.I.C.C. Monograph No 2, 1967.

DE REUCK, A. V. S., KNIGHT, J., eds.: Ciba Foundation Symposium on Cell Differentiation. London: J. & A. Churchill, Ltd. 1967.

DEWYS, W. D., HUMPHREYS, S. R., GOLDIN, A.: Studies on therapeutic effectiveness of drugs, etc. Cancer Chemother. Rep. 52, 229—242 (1968).

DJERASSI, I., FARBER, S.: Control and prevention of hemorrhage; platelet transfusion. Cancer Res. 25, 1499—1503 (1965).

DRUCKREY, H., PRESSMANN, R., IVANKOVICZ, S., SCHMÄHL, K.: Organotrope carcinogene Wirkungen bei 65 verschiedenen N-Nitroso-Verbindungen an BD-Ratten. Z. Krebsforsch. 69, 103—201 (1967).

ELION, G. B., NATHAN, H., BIEBER, S., RUNDLES, R. W., HITCHINGS, G. H.: Potentiation by inhibition of drug degradation: 6-substituted purines and xanthine oxidase. Biochem. Pharmacol. 12, 86—93 (1963). See also Symposium on Allopurinol (ed. J. T. SCOTT). Ann. rheum. Dis. 25, No 6 (Suppl.), 599—718 (1966).

ELSON, L. A., JARMAN, M., ROSS, W. C. J.: Toxicity, haematological effects and anti-tumour activity of epoxides derived from disubstituted hexitols. Mode of action of mannitol myleran and dibromomannitol. Europ. J. Cancer 4, 617—625 (1968).

ENDICOTT, K. M.: U.S. strategy against cancer. New Scientist 856—867 (Sept. 1965).

FARBER, S.: Clinical and biological studies with actinomycin. In: Amino Acids and Peptides. Ciba Foundation Symposium (eds. G. E. W. WOLSTENHOLME and C. M. O'CONNOR), p. 138—148. London: J. & A. Churchill Ltd. 1958.

— Remarks during symposium on obstacles to the control of acute leukemia. Cancer Res. 25, 1472 and at closing session (1965).

— Chemotherapy in the treatment of leukemia and Wilm's tumor. J. Amer. med. Ass. 198, 826—836 (1966).

— Concluding remarks in symposium on acute leukemia and Burkitt's tumor. Cancer Res. 27, 2658—2660 (1967).

— MITUS, A. T.: Role of actinomycin D (Dactinomycin). Pediatric oncology. In: Actinomycin (ed. S. A. WAKSMAN), p. 137—145. New York: Intersc. Publ. 1968.

FOLEY, G. E., BARELL, E. F., ADAMS, R. A., LAZARUS, H.: Nutritional requirements of human leukemic cells: cystine requirements of diploid cell lines and their heteroploid variants. J. exp. Cell Res. 57, 129—133 (1969).

FOX, B. W.: The sensitivity of Yoshida sarcoma to methylene dimethanesulphonate. Int. J. Cancer 4, 54—60 (1969).

— JACKSON, H.: In vivo effects of methylene dimethane sulphonate in proliferating cell systems. Brit. J. Pharmacol. 24, 24—28 (1965).

FREEMAN, R. G., KNOX, J. M.: Treatment of skin cancer. In: Recent results in cancer research, vol. 11, p. 19ff. Berlin-Heidelberg- New York: Springer 1967.

FREI III, E.: Summary of informal discussion on clinical obstacles to the control of acute leukemia. Cancer Res. 25, 1570 (1965).

FREI III, E.: New agents and combination therapy. In: Z. B. PAPANASTASSIOU Memorial Symposium on Current Trends in Cancer Chemotherapy. Med. Chem. Group, N. F. Section, Amer. Chem. Soc., May 1969.

FRENCH, F. A., FREEDLANDER, B. L.: Chemotherapy studies on transplanted mouse tumors. Cancer Res. **20**, 505—538 (1960).

FULLERTON, W. W., BOGGUST, W. A., O'MEARA, R. A. Q.: Antithromboplastic activities of fatty acids. J. clin. Path. **20**, 624—628 (1967).

GARATTINI, S., CONNORS, T. A., ALBERT, A.: Training course in cancer chemotherapy, Comm. on. Exp. Chemother. of the Comm. on Exptl. Oncol., U.I.C.C. in conj. with Polish Acad. Sciences, Warsaw, Okt. 1968a.

— AMOS, D. B., GOLDIN, A., MILDER, J., RALL, D. P., MANDEL, H. G.: Conference on the Critical Evaluation of Cancer Chemotherapy. Cherry Hill, New Jersey, April 1969.

— DONELLI, M. G., MORASCA, L., RAINISIO, C., ROSSO, R.: Attempts to establish new experimental methods to study antitumural drugs. Acta Genet. med. (Roma) **17**, 60—66 (1968b).

GERICKE, D.: DL-Glycerinaldehyd eine cytostatische Substanz? Krebsforschungslabor Farbwerke Hoechst. Abstr. 577, Int. Chemother. Kongr. Wien: Med. Acad. 1967.

— Aspekte der Immuntherapie Maligner Tumoren. Med. Klin. **63**, 1143—1146 (1968).

— CONTZEN, H.: Weitere Untersuchungen zur Onkolyse. Abstr. 1081 Int. Chemother. Kongr. Wien: Med. Akad. 1967.

— ENGELBERT, K.: Onkolysis by *Clostridia* II. Experiments on a variety of *Clostridia* in combination with heavy metal. Cancer Res. **24**, 217—221, see parts I, III, etc. same volume (1964).

— SCHÜTZE. E.: Versuche zur Beeinflussung des Tumorenwachstums durch Mykoplasmen. Zbl. Bakt., I. Abt. Orig. **210**, 212—216 (1969).

GOLD, P.: Tumor-specific antibody production in digestive system of cancer patients. Symposium on Immunological Aspects of Malignant Disease, 59th Ann. Mtg. Amer. Ass. Cancer Res. Atlantic City, N. J., 12th April 1968.

GOLDENBERG, I. S., MORIN, J. E., CAHOW, C. E.: Hormonal therapy for metastatic breast carcinoma in women VII. 6α-chloro-16α-methylpregn-4-ene-3,20-dione. Cancer Chemother. Rep. **50**, 327—329 (1966).

GOLDIN, A., KAZIWARA, K., KINOSITA, R., YAMAMURA, Y. (eds.): Cancer Chemotherapy. Jap. Cancer Ass., Gann Monograph **2**, Tokyo: Maruzen Co. Ltd. (1967).

— SERPICK, A. A., MANTEL, N.: Experimental screening procedures and clinical predictability value. Cancer Chemother. Rep. **50**, 173—218 (1966).

— WOOD, H. B., JR., ENGLE, R. R.: Relation of structure of purine and pyrimidine nucleosides to antitumor activity. Cancer Chemother. Rep. (Suppl.) **1**, 1—268 (1968).

GORER, P. A.: Isoantigens of malignant cells. In: Biol. appr. to cancer chemother. (ed. R. C. J. HARRIS), p. 219—230. London: Academic Press 1961.

GREENWALD, E. S.: Cancer chemotherapy. London: Heinemann 1967.

HADDOW, A.: Tumour immunology. Brit. med. Bull. **21**, 133—139 (1965).

— ALEXANDER, P.: An immunological method of increasing the sensitivity of primary sarcomas to local irradiation with X-rays. Lancet **1964I**, 452—457.

HALL, T. C. (in place of S. FARBER): Member of the cancer chemotherapy panel. Abstr. of Papers 12, IXth Int. Cancer Congr. Tokyo 1966.

— KESSEL, D., GODSILL, A., ROBERTS, D.: Uridine phosphorylation, an overlooked pathway?; 5-fluorouridine, a neglected drug? Proc. Amer. Ass. Cancer Res. **9**, 27, 103 (1968).

HAMILTON FAIRLEY, G., SIMISTER, J. M., eds.: Cyclophosphamide. Bristol: J. Wright & Sons, Ltd. 1964.

HANCOCK, P. E. T.: T. Hodgkin. The Fitz Patrick Lecture. J. roy. Coll. Phycns (Lond.) **2**, 404—421 (1968).

HANDSCHUHMACHER, R. E.: Discussion remark (not recorded) during Symposium of Amer. Cancer Soc., Conference on Obstacles to the Control of Acute Leukemia 1965.

HARRAP, K. R., JACKSON, R. C.: Some biochemical aspects of leukaemias; leucocyte glutathione metabolism in chronic granulocytic leukemia. Europ. J. Cancer 5, 61—67 (1969).

HARRIS, R. J. C., ed.: Proc. IXth. Int. Cancer Congr., UICC Monograph Ser. Panel 4. Berlin-Heidelberg-New York: Springer 1967.

HARTMANN, J. R., Chm: Clinical study of fluorometholone in acute leukemia in children. Cancer Chemother. Rep. 50, 339—345 (1966).

HEIDELBERGER, C.: Fluorinated pyrimidines — biochemically and clinically useful antimetabolites. In: Molekulare Biologie des malignen Wachstums, p. 156—176 (HOLZER and HOLLDORF, eds.). Berlin-Heidelberg-New York: Springer 1966.

HELLMAN, K.: Org., Symposium on Experimental Aspects of Cancer Chemotherapy (tape recordings). Imperial Cancer Res. Fund, London 1968.

HERTZ, R., LEWIS, J., LIPSETT, M. B.: Five years' experience with chemotherapy of metastatic choriocarcinoma and related trophoblastic tumors in women. Amer. J. Obstet. Gynaecol. 82, 631—640 (1961).

HILL, J. M., ROBERTS, J., LOEB, E., KHAN, A., McLELLAN, A., HILL, R. W.: L-Asparaginase therapy for leukemia and malignant neoplasms. J. Amer. Med. Assoc. 202, 882—888 (1967).

HOELZEL, F., HEBBELIN, H., HOECKE, E., MAAS, H.: Die Penetration und die intrazelluläre Verteilung von Trenimon, p-Benzochinon, Actinomycin und TEM bei in vitro-Untersuchungen mit EHRLICH Ascitestumor Zellen. Z. Krebsforsch. 70, 74—94 (1968).

HOLZER, H.: Wirkungsmechanismus der Zytostatika. Dtsch. med. J. 12, 312—317 (1961).

— Intrazelluläre Regulation des Stoffwechsels. Naturwissenschaften 50, 260—269 (1963); see also: Diskussion (Leitung: H. HOLZER). In: Molekulare Biologie des malignen Wachstums, p. 74—78. Berlin-Heidelberg-New York: Springer 1964.

HONDIUS BOLDINGH, W., LAURENCE, E. B.: Extraction, purification and preliminary characterisation of the epidermal chalone: a tissue specific mitotic inhibitor obtained from vertebrate skin. Europ. J. Biochem. 5, 191 (1968).

HOROWITZ, B., MADRAS, B., MEISTER, A., OLD, L. J., OLD, E. A.: L-Asparagine synthetase activities of mouse leukemias. Proc. Amer. Ass. Cancer Res. 9, 33, 127 (1968).

HÓRVATH, I. P., INSTITORIS, L.: Influence of the chemical structure on the biological tendency of cytostatic compounds related to dibromomannitol. Arzneimittel-Forsch. 17, 149—155 (1967).

HOWARD, J. P., CEVIK, N., MURPHY, M. L.: Cytosine arabinoside in acute leukemia in children. Cancer Chemother. Rep. 50, 287—291 (1966).

ISRAEL, M., MADDOCK, C. L., MODEST, E. J.: Experimental antitumor activity of synthetic polyamines. Abstr. Papers, SO 555, p. 320. IXth Int. Cancer Congr. Tokyo 1966.

IVERSEN, O. H.: Chalones of the skin. In: Joint Symposium Ciba Foundation and Wellcome Trust on Homeostatic Regulators, p. 29—56. London: J. & A. Churchill, Ltd. 1969.

JAQUES, P. J.: Lysosomes and homeostatic regulation. In: Joint Ciba Foundation and Wellcome Trust Symposium on Homeostatic Regulators (eds. G. E. W. WOLSTENHOLME and J. KNIGHT), p. 180—196. London: J. & A. Churchill, Ltd. 1969.

JARMAN, M., Ross, W. C. J.: The formation of epoxides from substituted hexitols. Carbohydr. Res. 9, 1 139—147 (1969a).

— — 4-substituted nicotinic acids and nicotinamides. Pt. II. The preparation of 4-methyl-nicotinamide riboside. J. chem. Soc. 199—203 (1969b).

JOHNSON, J. M., BERGEL, F.: Biological alkylating agents. In: HOCHSTER and QUASTEL (eds.), Metabolic inhibitors, vol. II, p. 161—192. New York: Academic Press 1963.

JOHNSON, R. O., BISEL, H., ANDREWS, N., WILSON, W., and others: Phase I clinical study of 6α-methylpregn-4-ene-3,11,20-trione. Cancer Chemother. Rep. **50**, 671—673 (1966).

KAUNG, D. T., WALSH, W. S., SBAR, S., PATNO, M. E.: Hydroxyureas in therapy for nonresectable cancer of the lung. Cancer Chemother. Rep. **52**, 271—274 (1968).

KEAST, D., GRIEVE, G. M., STANLEY, N. F.: Chemotherapy of reovirus-induced murine lymphoma 2731/L. Cancer in Africa **23**, 1 (1968).

KENNEDY, B. J.: Effect of massive doses of estradiol. undecylate in advanced breast cancer. Cancer Chemother. Rep. **51**, 491—495 (1967).

KESSEL, D., HALL, T. C., ROSENTHAL, D.: Uptake and phosphorylation of cytosine arabinoside by normal human blood cells in vitro. Cancer Res. **29**, 459—463 (1969).

KLEIN, E., FARBER, S., DJERASSI, I.: Control and prevention of hemorrhage; platelet separation. Cancer Res. **25**, 1504—1509 (1965).

KLEIN, G., SJÖGREN, H. O.: Humoral and cellular factors in homograft and isograft immunity against sarcoma cells. Cancer Res. **20**, 452—461 (1960).

— Discussion remark. In: Symposium on Biological Approaches to Cancer Chemotherapy (ed. R. C. J. HARRIS), p. 230. London: Academic Press 1961 a.

— Population changes in drug resistance in tumours. In: Symposium on Biological Approaches to Cancer Chemotherapy (ed. R. C. J. HARRIS), p. 201—217. London: Academic Press 1961 b.

— Humoral and cell-mediated mechanisms for host defense in tumor immunity. In: Viruses inducing Cancer, Implication for Therapy (W. J. BURDETT, ed.), p. 323—349. Salt Lake City: Univ. of Utah Press 1966.

— KLEIN, E., CLIFFORD, P.: Search for host defences in Burkitt lymphoma: Membrane fluorescence tests on biopsies and tissue culture lines. Cancer Res. **27**, 2510—2520 (1967).

KLINE, I.: Chemotherapy of leukemia L 1210 1-β-D-Arabinofuranosylcytosin Hydrochloride. Cancer Res. **26**, 853—859; 1930—1937 (1966).

KOLDOVSKY, P.: Tumor specific transplantation antigen. In: Recent result in cancer research, vol. 22. Berlin-Heidelberg-New York: Springer 1969.

KRAKOFF, I. H., CLIFFORD, G. O.: Clinical studies of N-hydroxyurethan. Cancer Chemother. Rep. **52**, 635—639 (1968).

LARIONOV, L. F.: Cancer chemotherapy. London and New York: Pergamon Press Ltd., 1965.

LAW, L. W.: Presidential address, "Studies of the significance of tumor antigens in the induction and repression of neoplastic diseases". 59th Ann. Mtg. Amer. Ass. Cancer Res. Cancer Res. **29**, 1—22 (1969).

LAWLEY, P. D.: Mechanism of action of alkylating agents: comparison with other cytotoxic, mutagenic and carcinogenic agents. In: Molekulare Biologie des malignen Wachstums, p. 126—141. See also Diskussion ibid., p. 150—155. Berlin-Heidelberg-New York: Springer 1966.

LAZARUS, H., MCCOY, T. A., FARBER, S., FOLEY, G. E.: Nutritional requirements of human leukemic cells. Asparagine requirements and the effects of L-asparaginase. J. exp. Cell Res. **57**, 134—138 (1969).

LEFKOWITZ, E., PAPAC, R. J., BERTINO, J. R.: Head and neck cancer. Cancer Chemother. Rep. **51**, 305—311 (1967).

LEWERENZ, H. J.: Zur Wirkung von DL-Glyceraldehyd auf permanente Kulturen menschlicher Zellen. Z. Krebsforsch. **69**, 260—269 (1967).

LI, C. P., PRESCOTT, B., USDIN, F., EDDY, B. E., MARTINO, E. C.: Local chemotherapy of neoplasm in hamsters with clam extract. Abstr. Papers, SO 583, IXth Int. Cancer congr. Tokyo 1966.

LISS, E., PALME, G., OEFF, K.: In vivo-Versuche über die Wirkung des Trenimons I. Z. Krebsforsch. **71**, 89—98 (1968).

Loo, T. L.: Effect of chemotherapy on kinetics of leukemic cells. Cancer Chemother. Rep. **50**, 296—297 (1966).

Loveless, A.: Possible relevance of O-6 alkylation of deoxyguanosine to the mutagenicity and carcinogenicity of nitrosamines and nitrosamides. Nature (Lond.) **223**, 206—207 (1969).

Lovesey, A. C., Ross, W. C. J.: Potential coenzyme inhibitors. Pt. II. Reduction of 4-methylnicotinamide derivatives by sodium dithionate and sodium borohydride. J. chem. Soc. 192—195 (1969).

Maeda, H., Kumagai, K., Ishida, N.: Characterization of neocarcinostatin. J. Antibiot. (Tokyo), Ser. A **19**, 253—259 (1966).

Mandel, H. G.: Symposium on critical evaluation of cancer chemotherapy. Bull. Cancer **7**, 2 (1969).

Manuila, L, Moles, S., Rentchnick, P. (eds.): New trends in the treatment of cancer. In: Recent results in cancer research, vol. 8. Berlin-Heidelberg-New York: Springer 1967.

Marsden, H. B., Steward, J. K. (eds.): Tumours in children, Wilm's tumour, p. 225—249. In: Recent results in cancer research, vol. 13. Berlin-Heidelberg-New York: Springer 1968.

Mashburn, L. T., Wriston, J. C.: Tumor-inhibitory effect of L-asparaginase from *Escherichia coli*. Arch. Biochem. **105**, 450—452 (1964).

Mathé, G., ed.: Scientific basis of cancer chemotherapy. In: Recent results in cancer research, vol. 21. Berlin-Heidelberg-New York: Springer 1969a.

— Org. Dir. U. I. C. C., Advanced Course on Tumour Antigens at the Institut de Cancerologie et D'Immunogenétique, Villejuif, France, December 1969b.

— Amiel, J. L., Schwarzenberg, C., Cattan, A., Schneider, M.: Adoptive immunotherapy of acute leukemia, experimental and clinical results. Cancer Res. **25**, 1525—1531 (1965).

— Schwarzenberg, L., Amiel, J. L., Schneider, M., Cattan, A., Schlumberger, J. R.: The role of immunology in the treatment of leukemia and hematosarcomas. Cancer Res. **27**, 2542—2553 (1967).

Mauger, A. B., Wade, R.: The synthesis of actinomycin analogues. Pt. II. Actinocylgramicidin S. J. Chem. Soc. 1406—1408 (1966).

Meienhofer, J.: A total synthesis of actinomycin. Experientia (Basel) **24**, 776—777 (1968).

Mihich, E.: Combined effects of chemotherapy and immunology against leukemia L 1210 DBA/2 mice. Cancer Res. **29**, 848—855 (1969).

— Hakala, M. T.: Differences between the antitumor action of 4,4'-Di-acetyl-diphenyl-urea-bis-guanylhydrazone and methylglyoxal-bis-guanylhydrazone. Proc. Amer. Ass. Cancer Res. **9**, 48, 190 (1968).

Milder, J. W.: Introductory remarks about symposia in 1964, 1965 and 1967. Cancer Res. **27**, 2419 (1967).

Mitchell, J. S. (ed.): The treatment of cancer. Cambridge, Engl.: Cambridge University Press 1965.

Modest, E. J., Foley, G. E., Farber, S.: Polypeptides and proteins as inhibitors. In: Metabolic inhibitors (R. M. Hochster and J. H. Quastel, eds.), p. 75—129. New York and London: Academic Press 1963.

— Sengupta, S. K., Tinter, S. K., Trites, D. H.: Laboratories of Organic Chemistry, Children's Cancer Res. Found., Boston, Mass.: Private Communication 1969.

Mohr, O., Althoff, J., Kinzel, V., Süss, R., Volm, M.: Melanoma regression induced by "chalone": a new tumour inhibitory principle acting *in vivo*. Nature (Lond.) **220**, 138—139 (1968).

Monod, J., Changeux, J. P., Jacob, F.: Allosteric proteins and cellular control systems. J. molec. Biol. **6**, 306—329 (1963).

Mühlbock, O. (Dir.): Techniques with experimental animals in cancer research. World Health Org. and Int. Agency for Res. on Cancer. Amsterdam, May 1969.

Nass, M. M. K.: Mitochondrial DNA, advances, problems and goals. Science 165, 25—35 (1969).

Nathanson, L., Hall, T. C., Dederick, M. M., Yount, W., Miller, S.: Initial pharmacological studies of three types of combination chemotherapy. Cancer Chemother. Rep. 50, 259—264 (1966).

Nossal, G. J. V.: The impact of immunology on the pediatrics of the future (in press). Rep. Proc. Children's Hosp. Med. Center, 100th Anniv. Celebr. 1969.

O'Bryan, R. M., Talley, R. W.: Preliminary pharmacology dose-response studies of 7α-methyl-19-nortestosterone in patients. Cancer Chemother. Rep. 50, 335—338 (1966).

Ochoa, M., Jr., Hirschberg, E.: Alkylating agents in experimental chemotherapy, vol. V., pt. 1. New York: Academic Press Inc. 1967.

Oettgen, H. F.: Inhibition of leukemias in man by L-asparaginase. Cancer Res. 26, 2619—2631 (1967).

Ohnuma, T., Bergel, F., Bray, R. C.: Enzymes in cancer: asparaginase from chicken liver. Biochem. J. 103, 238—245 (1966).

Okamoto, H., Shoin, S., Minami, M., Koshimara, S., Shimizu, R.: A Development in the Study of Anticancer Activity of Hemolytic Streptococci, SO 580, Abstr. Papers, IXth Int. Cancer Congr. Tokyo 1966.

Ono, Y., Ito, Y., Maeda, H., Ishida, N.: Mode of action of neocarcinostatin: requirement of protein synthesis for neocarzinostatin-mediated DNA degradation in Sarcina lutea. Biochem. Acta 155, 616—618 (1968).

Papaiannou, A. N., Volk, H.: Massive doses of Δ'-testostololactone for advanced breast cancer. Chemother. 50, 323—326 (1966).

Plattner, P. A. (ed.): Chemotherapy of cancer. Amsterdam: Elsevier Publ. Co. 1964.

Porter, R., Wiltshaw, E. (eds.): Methotrexate in the treatment of cancer. Bristol: J. Wright & Sons, Ltd. 1962.

Pressman, D. (Chm.): Symposium on Immunological Aspects of Malignant Disease. 59th Ann. Mtg. Amer. Ass. Cancer Res., Atlantic City, 12th April 1968.

Rall, D. P., Homan, E. R.: New concepts in cancer chemotherapy. Cancer Chemother. Rep. 51, 247—251 (1967).

Regelson, W.: Antimitotic activity of polyanions. In: Adv. in chemotherap. (A. Goldin, F. Hawking and R. S. Schnitzer, eds.), vol. 3, p. 303—370. New York: Academic Press 1968.

— Holland, J. F., Talley, R. W.: Clinical pharmacological study of kethoxal bis-(thiosemicarbazone) in advanced cancer. Cancer Chemother. Rep. 51, 171—177 (1967).

Roberts, D., Hall, T. C.: Enzyme activity and deoxynucleoside utilization of leukemic leukocytes in relation to drug therapy and resistance. Cancer Res. 29, 166—173 (1969).

Rosenoer, V. M., Whisson, M. E.: A transplantable plasma cell tumour in the study of carcinostatic agents. Biochem. Pharmacol. 13, 589—602 (1964).

Ross, G. T., Stellbach, L., Hertz, L.: Actinomycin D in the treatment of methotrexate-resistant trophoblastic diseases in women. Cancer Res. 22, 1015—1017 (1962).

Ross, W. C. J.: Biological alkylating agents. London: Butterworth 1962.

— Anti-coenzymes. Brit. Emp. Cancer Camp., Ann. Rep. 42, 81 (1964).

— The preparation of some 4-substituted nicotinic acids and nicotinamides. J. chem. Soc. 1816—1821 (1966).

— Some alkylating derivatives of nicotinic acids. J. med. Chem. 10, 257 (1967a).

— Some 6-substituted nicotinamides. Biochem. Pharmacol. 16, 675 (1967b).

— Jarman, M.: Nicotinamide ribosides. Brit. Emp. Cancer Camp., Ann. Rep. 44, 14 1966); 45, 38 (1967).

Rosso, R., Donelli, M. G., Franchi, G., Garattini, S.: Effect of Triton WR 1339 on cancer dissemination and metastasis. Europ. J. Cancer 5, 77—78 (1969).

Rusconi, A., Fronzo, G. di, Dimarco, A.: Distribution of tritiated daunomycin in normal rats. Cancer Chemother. Rep. 52, 331—337 (1968).

Rytömaa, T., Kiviniemi, K.: Control of cell production in rat chloroleukaemia by means of the granulocytic chalone. Nature (Lond.) 220, 136—138 (1968a).

— — Control of DNA duplication in rat chloroleukaemia by means of the granulocytic chalone. Europ. J. Cancer 4, 595—606 (1968b).

— — Control of granulocyte production. Cell Tissue Kinet. 1, 329—340, 341—350 (1968c).

Sachs, L.: An analysis of the mechanism of carcinogenesis by polyoma virus, hydrocarbons and X-irradiation. In: Molekulare Biologie des malignen Wachstums (Holzer and Holldorf, eds.). Berlin-Heidelberg-New York: Springer 1966.

Sadler, P. W.: Antiviral chemotherapy with isatin-β-thiosemicarbazone and its derivatives. In: Symposium on Drugs, Parasites and Hosts (L. G. Goodwin and R. H. Nimmo-Smith, eds.), p. 286—293. London: J. & A. Churchill, Ltd. 1962.

Sahasrabudhe, M. B.: A new approach to chemotherapy and radiotherapy. J. Sci. Ind. Res. 26, 243—248 (1967).

Sandford, K. K., Earle, W. R., Likely, G. D.: The growth in vitro of single isolated tissue cells. J. nat. Cancer Inst. 9, 229 (1948).

Scanlon, E. T., Hawkins, R. A., Fox, W. W., Smith, W. S.: Fatal homotransplanted melanoma. Cancer (Philad.) 18, 782—794 (1965).

Schmidt, L. H., Fradkin, R., Sullivan, R., Flowers, A.: Comparative pharmacology of alkylating agents. Cancer Chemother. Rep., pts. I, II, and III, Suppl. 2 (Jan. 1965).

Shimkin, M. B. (ed.): Proc. Amer. Ass. Cancer Res. 9, 1—83 (1968).

Simnet, J. D., Fisher, J. M., Heppelstone, A. G.: Tissue specific inhibition of lung alveolar cell mitosis in organ culture. Nature (Lond.) 223, 944—946 (1969).

Skipper, H. E.: The effects of chemotherapy on the kinetics of leukemic cell behaviour. Cancer Res. 25, 1544—1550; Summary of informal discussion on the effects of chemotherapy on the kinetics of leucemic cell behaviour. Cancer Res. 25, 1553—1554 (1965).

— Schabel, F. M., Jr., Wilcox, W. S.: On the criteria and kinetics associated with "curability" of experimental leukemias. Cancer Chemoth. Rep. 35, 3—111 (1964).

Smithers, D. W.: On the nature of neoplasia in man. Edinburgh and London: E. & S. Livingstone 1964.

Southam, C. M.: Evidence of immunological reactions to autochthonous cancer in man. Europ. J. Cancer 1, 173—181 (1965).

— Summary: Immunology of acute leukemia and Burkitt's tumor. Cancer Res. 27, 2554—2556 (1967).

— Co-existence of allogeneic tumour growth and homograft immunity in man. Europ. J. Cancer 4, 507—511 (1968).

Squibb institute for medical research, New Brunswick, N. J., Proc. of Conference on L-Asparaginase, 14th March 1968.

Stanley, W. M.: Vaccines for viral diseases of children, past, present and future. Rep. Proc. Children's Hosp. Med. Center. 100th Anniv. Celebr. (in press) 1969.

Stedman's medical dictionary, 20th ed. London: Baillière, Tindall & Cox 1966.

Stock, J. A.: Antimetabolites. In: Experimental chemotherapy, vol. IV, p. 79—237 (Schnitzer and Hawkings, eds.). New York: Academic Press, Inc. 1966a.

— Antitumour antibiotics. In: Experimental chemotherapy (R. J. Schnitzer and F. Hawking, eds.), vol. IV, p. 241—377. New York: Academic Press, Inc. 1966b.

— Other antitumor agents. In: Experimental chemotherapy, vol. V, pt. II, p. 334—416. New York: Academic Press 1967a.

— Some thoughts on cancer chemotherapy and the future. In: Experimental Chemotherapy, vol. V, pt. II, p. 451—459. New York: Academic Press Inc. 1967b.

Stoker, M.: Regulation of growth and orientation in hamster cells transformed by polyoma virus. Virology 24, 165—174 (1964).

STOKER, M., MACPHERSON, I.: Studies on transformation of hamster cells by polyoma virus *in vitro*. Virology 14, 359—370 (1961).

SUGIHARA, Y., ARAKI, F.: Carcinostatic substances prepared from bovine liver. Abstr. Papers SO 589, IXth Int. Cancer Congr., Tokyo 1966.

SUGUIRA, K., BROWN, G. B.: Antitumor activity of purine N-oxides. Abstr. Paper p. 317 (SO 549). IXth Int. Cancer Congr., Tokyo 1966.

— — Purine N-oxides. XIX. On the oncogenic derivatives of guanine and xanthine and a nononcogenic isomer of xanthine N-oxide. Cancer Res. 27, 925—931 (1967).

SULLIVAN, M. P. (Chrm.): Symposium on Vincristine. Memphis, Tennessee. Cancer Chemother. Rep. 52, 453—533 (1968). See also: Antitumoral effects of *Vinca Rosea* alkaloids. In: Proc. 1st Symp. Europ. Chemother. Group, Paris, June 1965. Int. Congr. Ser. No 106 (1965).

SZEKERKE, M., WADE, R., BERGEL, F.: Cytoactive amino acids and peptides. Pt. XIV. Poly-acid copoly-amino acid derivatives of melphalan. J. chem. Soc. 1792—1795 (1968).

SZENT GYÖRGYI, A.: Bioelectronics. Science 161, 988—990 (1968).

TAGNON, H. J.: Note. Europ. J. Cancer 4, 507 (1968).

TELLER, M., SUGUIRA, K., PARHAM, J. C., BROWN, G. B.: Oncogenicity of purine N-oxide derivatives. Proc. Amer. Ass. Cancer Res. 9, 70 (277) (1968).

THOMAS, H. J., MONTGOMERY, J. A.: Complex esters of thioinosinic-(5') acid. J. med. pharm. Chem. 5, 24—32 (1962).

TIMMIS, G. M., WILLIAMS, D. C.: Chemotherapy of cancer, the antimetabolic approach. London: Butterworths 1967.

UMEZAWA, H.: The distribution of H-bleomycin in mouse tissue. J. Antibiotics (Tokyo) 21, 638—642 (1968).

VENDITTI, J. M., ABBOTT, B. J., DIMARCO, A., GOLDIN, A.: Effectiveness of daunomycin against experimental tumors. Cancer Chemother. Rep. 50, 659—665 (1966).

VESELY, J.: Cancerostatic effects of recently synthesized thymine mustards and their fluoro derivatives. SO 551, Abstr. Papers, IXth Int. Cancer Congr., Tokyo 1966.

WADE, R.: Hormones. In: Experimental Chemotherapy, vol. V, p. 133—331 (SCHNITZER and HAWKINGS, eds.). New York: Academic Press 1967.

— WHISSON, M. E., SZEKERKE, M.: Some serum protein nitrogen mustard complexes with high chemotherapeutic selectivity. Nature (Lond.) 215, 1303—11304 (1967).

WAKSMAN, S. A.: The actinomycetes and their antibiotics. Advanc. appl. Microbiol. 5, 235—315 (1963).

— FURNESS, F. N.: The actinomycins and their importance in the treatment of tumors in animals and man. Ann. N. Y. Acad. Sci. 89, 283—486 (1960).

WARBURG, O.: Über die Ursache des Krebses, p. 1. In: Molekulare Biologie des malignen Wachstums (HOLZER and HOLLDORF, eds.). Berlin-Heidelberg-New York: Springer 1966.

— GAWEHN, K., GEISSLER, A. W., LORENZ, S.: Über die Heilung von Mäuse-Ascites-Krebs durch D- and L-Glycerinaldehyd. Z. klin. Chem. 1, 175—177 (1963).

WEITZEL, G. F., SCHNEIDER, D., KUMMER, D., OCHS, H.: Cytostatischer Wirkungsmechanismus von Natulan. Z. Krebsforsch. 70, 354—365 (1968).

WHISSON, M. E.: The interaction of tumour and embryonic tissue *in vivo*. In: Symposium Ciba Foundation on Cell Differentiation (A. V. S. DE REUCK and J. KNIGHT, eds.), p. 219—231. London: J. & A. Churchill, Ltd. 1967.

WOLFF, E., WOLFF (Mme), E.: Factors of growth and maintenance of tumours as organised structures *in vitro*. In: Ciba Foundation Symposium on Cell Differentiation (A. V. S. DE REUCK and J. KNIGHT, eds.), p. 208—215. London: J. & A. Churchill, Ltd. 1967.

WOLSTENHOLME, G. E. W., KNIGHT, J. (eds.): Homeostatic regulators, Joint Ciba Foundation-Wellcome Trust Symposium, Jan. 1969 (in press). London: J. & A. Churchill, Ltd. 1969.

WOODRUFF, H. B., MILLER, I. M.: Antibiotics in metabolic inhibitors (B. M. HOCHSTER and J. H. QUASTEL, eds.), vol. II, p. 23—51. New York and London: Academic Press 1963.

YOSHIDA, T.: org. chrm., Publications IXth Int. Cancer Congr., Tokyo 1966.

ZILBER, L. A.: An immunological approach to tumour growth control. In: Symposium on Biological Approaches to Cancer Chemotherapy (R. C. J. HARRIS, ed.), p. 231—243. London: Academic Press 1961.

ZUBROD, C. G.: Quantitative concepts in the clinical study of drugs. In: Advances in Chemotherapy, vol. 1, p. 9—34. New York and London: Academic Press 1964.

— Treatment of the acute leukemias. Cancer Res. 27, 2557—2560 (1967).

— SCHEPARTZ, S., LEITER, J., ENDICOTT, K. M., CARRESE, L. M., BAKER, C. G.: Program of the National Cancer Institute; history, analysis and plans. Cancer Chemother. Rep. 50, 349—540 (1966).

Namenverzeichnis

Die in Klammern stehenden Ziffern beziehen sich auf die Nummern der Zitate innerhalb des laufenden Textes und der Literatur

Die gewöhnlich gesetzten Ziffern weisen auf die entsprechende Stelle im Text und die *kursiven* Seitenzahlen auf das Literaturverzeichnis hin

Hakala, M. T., s. Mihich, E. 114, *135*

Haldar, J., s. Bissett, G. W. 46, *60*

Hales, C. N., s. Randle, P. J. (106), 80, *88*

Hall, L. M. (64), 75, *87*

Hall, T. C. 97, 98, 107, *132*

— Kessel, D., Godsill, A., Roberts, D. 110, *132*

— s. Kessel, D. 110, *134*

— s. Nathanson, L. 128, *136*

— s. Roberts, D. 107, *136*

Hamberger, B., s. Norberg, K. A. 13, 21, *65*

Hamilton Fairley, G., Simister, J. M. 105, *132*

— s. Alexander, P. 118, 119, 120, *128*

Hancock, P. E., Hancock, P. E. T., Hodgkin, T. 127, *132*

Hancock, P. E. T., s. Hancock, P. E. 127, *132*

Handschuhmacher, R. E. 110, *133*

Harman, J. W., O'Hegarty, M. T., Byrnes, C. R. 21, *63*

Harmer, Iris Mary 1

Harrap, K. R., Jackson, R. C. 128, *133*

— s. Bergel, F. 104, 113, 114, *129*

Harris, J. S., s. Kaiser, I. H. 53, *64*

Harris, R. A., s. Allmann, D. W. (3), 75, *84*

Harris, R. J. C. 97, *133*

Hartmann, J. R. 125, *133*

Hasama, B. 39, *63*

Haugen, J. A., s. Adair, F. L. 53, *59*

Hawkins, D. F. 54, *63*

Hawkins, J. T., s. Bueding, E. 57, *60*

Hawkins, R. A., s. Scanlon, E. T. 119, *137*

Hebbelin, H., s. Hoelzel, F. 105, *133*

Heidelberger, C. 110, *133*

Hellmann, K. 98, 114, *133*

— s. Creighton, A. M. 114, *130*

Heppelstone, A. G., s. Simnet, J. D. 126, *137*

Hermansen, K. 49, 50, 51, 54, *63*

Hertz, L., s. Ross, G. T. 107, 127, *136*

Hertz, R., Lewis, J., Lipsett, M. B. 107, 127, *133*

Hill, J. M., Roberts, J., Loeb, E., Khan, A., McLellan, A., Hill, R. W. 113, *133*

Hill, R. W., s. Hill, J. M. 113, *133*

Hillarp, N. A., s. Carlsson, A. 45, *61*

— s. Falck, B. 12, *62*

Hinton, T., s. Fraenkel, G. (39), 69, *85*

Hirschberg, E., s. Ochoa, M. 101, *136*

Hitchings, G. H., s. Elion, G. B. 110, *131*

Hodgkin, T., s. Hancock, P. E. 127, *132*

Hoecke, E., s. Hoelzel, F. 105, *133*

Hoelzel, F., Hebbelin, H., Hoecke, E., Maas, H. 105, *133*

Holland, J. F., s. Regelson, W. 114, *136*

Holman, M. E. 23, 40, *63*

— s. Burnstock, G. 21, 23, 24, 39, 40, 41, 48, 55, 56, *60*

Holton, P., s. Cleugh, J. (15), 2, *4*

Holtrop, R. H., s. Fritz, I. B. (47), 75, *86*

Holtz, P., s. Greeff, K. 50, 51, 52, *63*

Holzer, H. (65), 71, *87*, 103, *133*

Homan, E. R., s. Rall, D. P. 98, *136*

Hondius Boldingh, W., Laurence, E. B. 125, *133*

Hopkins, T. F., Pincus, G. 44, *63*

Horowitz, B., Madras, B., Meister, A., Old, L. J., Old, E. A. 99, *133*

Horton, E. W. (20), 2, *4*

Horváth, I. P., Institoris, L. 104, *133*

Howard, J. P., Cevik, N., Murphy, M. L. 110, *133*

Hülsmann, W. C., Siliprandi, D., Ciman, M., Siliprandi (66), 80, *87*

— Wit-Peeters, E. M., Benkhuysen, C. (67), 80, *87*

Humphreys, S. R., s. Dewys, W. D. 98, *131*

Huntress, W. T., s. Bratzel, R. P. 101, *129*

Ikoku, C., s. Davidson, W. J. 50, *61*

Ingelman-Sunberg, A., s. Sandberg, F. 54, *66*

Institoris, L., s. Horváth, I. P. 104, *133*

Isaac, P. F., s. Pennefather, J. N. 36, *66*

Ishida, N., s. Maeda, H. 116, *135*

— s. Ono, Y. 116, *136*

Israel, M., Maddock, C. L., Modest, E. J. 115, *133*

Ito, Y., s. Ono, Y. 116, *136*

Ivankovicz, S., s. Druckrey, H. 103, *131*

Iversen, O. H. 124, 125, *133*

Ivy, A. C., s. Rudolph, L. 33, 34, 35, 41, *66*

Jackson, H., s. Fox, B. W. *131*

Jackson, R. C., s. Harrap, K. R. 128, *133*

Jacob, F., s. Monod, J. (96), 79, *88*, 107, *135*

Jacobowitz, D., Wallach, E. E. 14, 24, *63*

Jacobson, H. N., Nieves, O. 11, 12, 15, *64*

Jaeger, J. 21, *64*

Jaenicke, L., Lynen, F. (68), 71, *87*

Owens, Jr., A. H., s. Burke,
P. F. 127, *130*

Owman, C., Rosengren, E.,
Sjöberg, N. O. 14, 15,
16, 17, 25, 28, 29, *65*
— Sjöberg, N. O. 14, 15,
17, 18, 21, 24, *65*
— Sjöstrand, N. O. 18, *65*

Pallie, W., Corner, G. W.,
Weddell, G. 11, 12, *66*

Palme, G., s. Liss, E. 105,
134

Papa, S., s. Tager, J. M.
(126), *89*

Papac, R. J., s. Lefkowitz, E.
107, *134*

Papaiannou, A. N., Volk, H.
125, *136*

Parham, J. C., s. Teller, M.
111, *138*

Patno, M. E., s. Kaung, D. T.
114, *134*

Paton, D. M. 49, 52, *66*

Pauerstein, C. J., Woodruff,
J. D., Zachary, A. S.
54, *66*

Pearson, D. J., Tubbs, P. K.
(102, 103), 73, *88*
— s. Chase, J. F. A. (30),
72, *85*

Peart, W. S., Vogt, M. (23,
24), 2, *4*

Pennefather, J. N., Isaac,
P. F. 36, *66*

Phillips, G., s. Adrian, E. D.
30, *59*

Pincus, G., s. Brooks, J. R.
51, *60*
— s. Burdick, H. O. 42, *60*
— s. Hopkins, T. F. 44, *63*

Plattner, P. A. 95, 101,
102, 103, *136*

Plaut, G. W., Plaut, K. A.
(104), 75, *88*

Plaut, K. A., s. Plaut, G. W.
(104), 75, *88*

Pogell, B. M., s. Taketa, K.
(127), 79, *89*

Porter, R., Wiltshaw, E.
107, *136*

Pose, S. V., Cibils, L. A.,
Zuspan, F. P. 53, *66*
— s. Cibils, L. A. 53, *61*

Potter, L. T., s. Wurtman,
R. J. 27, *67*

Preiss, B., s. Meyer, H.
(95), 75, *88*

Prescott, B., s. Li, C. P.
117, *134*

Pressmann, D. 100, *136*

Pressmann, R., s. Druckrey,
H. 103, *131*

Prinzie s. De Somer 123

Purvis, J. L., Lowenstein,
J. M. (105), 76, *88*

Quagliariello, E., s. Tager,
J. M. (126), *89*

Rainisio, C., s. Garattini, S.
97, *132*

Rall, D. P., Homan, E. R.
98, *136*
— s. Garattini, S. 99, 127,
132

Randić Mirjana, Smith,
M. W. (36), 3, *5*

Randle, P. J., Garland,
P. B., Hales, C. N.,
Newsholme, E. A.
(106), 80, *88*

Regelson, W. 122, *136*
— Holland, J. F., Talley,
R. W. 114, *136*

Renold, A. E., s. Evans,
J. R. (37), 79, 80, *85*

Rentchnick, P., s. Manuila,
L. 92, *135*

Renzi, A. A., s. Gaunt, R.
44, *63*

Reynier, M. (107, 108),
69, 79, *88*

Reynolds, S. M. R. 7, 8,
27, 29, 30, 36, 43, 48,
66

Reynolds, S. R. M.,
s. Sauer, J. 33, 34, *66*

Richardson, K. C. 23, *66*

Ritzen, M., s. Norberg,
K. A. 21, *65*

Roberts, D., Hall, T. C.
107, *136*
— s. Hall, T. C. 110, *132*

Roberts, J., s. Hill, J. M.
113, *133*

Robinson, K., s. Daniel,
E. E. 47, *61*

Robinson, P. M., Burn-
stock, G. 21, 23, *60*

Rosengren, E., Sjöberg,
N. O. 14, 15, 17, 18,
19, 21, 28, 29, *66*
— s. Bertler, A. 12, 25, *60*
— s. Carlsson, A. 13, *61*
— s. Norberg, J. G. 21, *65*
— s. Owman, C. 14, 15,
16, 17, 25, 28, 29, *65*

Rosenoer, V. M., Whisson,
M. E. 104, *136*

Rosenthal, D., s. Kessel, D.
110, *134*

Ross, G. T., Stellbach, L.,
Hertz, L. 107, 127, *136*

Ross, R. B., s. Bratzel,
R. P. 101, *129*

Ross, W. C. J. 101, 112, *136*
— Jarman, M. 112, *136*
— s. Calvert, N. 105, *130*
— s. Connors, T. A. 111,
130
— s. Elson, L. A. 104, *131*
— s. Jarman, M. 105, 112,
133
— s. Lovesey, A. C. 112,
135

Rossi, C. R., Galzigna, L.,
Gibson, D. M. (109),
76, 80, *88*

Rossi, R., s. Garattini, S.
97, *132*

Rosso, R., Donelli, M. G.,
Franchi, G., Garattini, S.
122, *136*

Rotzsch, W., Lorenz, I.,
Strack, E. (110), 81, *88*

Rozenblum, C., s.
Broekhuysen, J. (28), *85*

Rudolph, L., Ivy, A. C.
33, 34, 35, 41, *66*

Rudzik, A., s. Cha, K. S.
28, *61*

Rudzik, A. D., Miller, J. W.
25, 26, 27, 51, *66*

Rüsse, M., Marshall, J. M.
36, 38, 44, *66*

Rumsay, P., s. Friedman,
A. D. (41), 80, *86*

Rundles, R. W., s. Elion,
G. B. 110, *131*

Rusconi, A., diFronzo, G.,
Dimarco, A. 116, *137*

Vogt, M., s. Vanov, S.　9, *67*
Volk, H., s. Papaiannou,
　A. N.　125, *136*
Volk, M. E., Millington,
　R. H., Weinhouse, S.
　(141), 75, *90*
Volm, M., s. Mohr, O.
　125, *135*

Wade, R.　124, *138*
— Whisson, M. E.,
　Szekerke, M.　104, *138*
— s. Bergel, F.　104, *129*
— s. Mauger, A. B.　116,
　135
— s. Szekerke, M.　104,
　138
Wagatsuma, T., s. Kumar,
　O.　55, 56, *64*
Waksman, S. A.　115, *138*
— Furness, F. N.　115, *138*
Walker　105
Wallach, E. E., s.
　Jacobowitz, D.　14, 24,
　63
Walsh, W. S., s. Kaung,
　D. T.　114, *134*
Wansbrough, H.　53, *67*
— s. Nakanishi, H.　41, *65*
Warburg, O.　94, 111, *138*
— Gawehn, K., Geissler,
　A.W., Lorenz, S.
　114, *138*
Webb, J. L., s. Montgomery,
　C. M.　(97), 75, *88*
Weddell, G., s. Pallie, W.
　11, 12, *66*
Weil-Malherbe, H., s.
　Gutman, Y.　27, *63*
Weinhouse, S., s. Volk,
　M. E.　(141), 75, *90*
Weiss, L., s. Wieland, O.
　(142, 143), 79, *90*
Weiss-Fogh, T., s. Krogh, A.
　(75), 75, *87*
Weitzel, G. F., Schneider,
　D., Kummer, D., Ochs,
　H.　105, *138*
West, G. B., s. Mann, M.
　33, 35, 37, *64*
Westermann, B., s.
　Wieland, O.　(143), 79,
　90

Whisson, M. E.　122, *138*
— s. Connors, T. A.
　104, *130*
— s. Rosenoer, V. M.
　104, *136*
— s. Wade, R.　104, *138*
Whitecross, S., s.
　Creighton, A. M.　114,
　130
Whittaker, V. P., s.
　Cleugh Joan　(37), 3, *5*
Wiederman, J., s.
　Freund, M.　47, *62*
Wieland, O., Weiss, L.
　(142), 79, *90*
— — Eger-Neufeldt, I.,
　Teinzer, A., Wester-
　mann, B.　(143), 79, *90*
Wilcox, W. S., s. Skipper,
　H. E.　*137*
Willems, J. L., Schaep-
　dryver, de A. F.　49, 50,
　67
Willgerodt, H., s.
　Thomitzek, W.-D.
　(132), 82, *89*
Williams, D.W., s. Timmis,
　G. M.　106, 107, 110,
　111, 122, *138*
Williamson, J. R., Krebs,
　H. A.　(144), 75, *90*
Wilson, W., s. Johnson,
　R. O.　125, *134*
Wiltshow, E., s. Porter, R.
　107, *136*
Winter, H., s. Thomitzek,
　W.-D.　(133), 71, *89*
Wit-Peeters, E. M., s.
　Hülsmann, W. C.　(67),
　80, *87*
Wolf, G.　(146), 70, 83, *90*
— Berger, C. R. A.　(145)
　70, 71, *90*
— s. Abdel-Kader, M. M.
　(1), 70, 71, *84*
— s. Khairallah, E. A.
　(73), 69, *87*
— s. Mehlman, M. A.
　(93, 94), 73, *88*
— s. Soderberg, J.　(114),
　73, *89*
Wolff, E., Wolff, E. (Mme)
　126, *138*

Wolffs, Ambrose　122
Wolstenhome, G. E. W.,
　Knight, J.　117, 123,
　138
Wood, C., s. Nakanishi, H.
　41, 42, *65*
Wood, H. B., Jr. s. Goldin,
　A.　109, 110, *132*
Woodruff, s. Baldwin　119
Woodruff, H. B., Miller,
　I. M.　115, *139*
Woodruff, J. D., s. Pauer-
　stein, C. J.　54, *66*
Woronkow, S., s. Gaffney,
　E.　28, *62*
Wriston, J. C., s. Mashburn,
　L. T.　112, *135*
Wurtman, R. J., Axelrod, J.,
　Kopin, I. J.　26, 28, *67*
— — Potter, L. T.　27,
　67
— Chu, W., Axelrod, J.
　25, *67*

Yamamura, Y., s. Goldin, A.
　96, *132*
Yates, D.W., s. Garland,
　P. B.　(58), 79, 81, *86*
— s. Shepherd, D.　(113),
　79, *88*
Yoshida, T.　96, 97, *139*
Yount, W., s. Nathanson, L.
　128, *136*
Yu, S.Y., s. Strength, D. R.
　(124, 125), 71, *89*
Yue, K. T. N., Fritz, I. B.
　(147), 81, 82, *90*
— s. Fritz, I. B.　(49, 50,
　52, 54a), 72, 73, 76, *86*,
　90

Zachary, A. S., s. Pauer-
　stein, C. J.　54, *66*
Zilber, L.A.　119, 121, *139*
Zubrod, C. G.　98, 99, 110,
　127, 128, *139*
— Schepartz, S., Leiter, J.,
　Endicott, K. M., Carrese,
　L. M., Baker, C. G.　98,
　128, *139*
Zuspan, F. P., s. Cibils, L.A.
　53, *61*
— s. Pose, S. V.　53, *66*

Sachverzeichnis